Medieval Animals on the Move

László Bartosiewicz • Alice M. Choyke
Editors

Medieval Animals on the Move

Between Body and Mind

palgrave
macmillan

Editors
László Bartosiewicz
Osteoarchaeological Research Laboratory
Stockholm University
Stockholm, Sweden

Alice M. Choyke
Department of Medieval Studies
Central European University
Vienna, Austria

ISBN 978-3-030-63887-0 ISBN 978-3-030-63888-7 (eBook)
https://doi.org/10.1007/978-3-030-63888-7

Cover illustration: Granger Historical Picture Archive / Alamy Stock Photo

This Palgrave Macmillan imprint is published by the registered company Springer Nature Switzerland AG.
The registered company address is: Gewerbestrasse 11, 6330 Cham, Switzerland

Preface

Medieval Animals on the Move: Between Body and Mind comprises a selection of peer-reviewed papers based on the 2017 conference of the Medieval Animal Database-Network (MAD), organized at the Department of Archaeology and Classical Studies of Stockholm University, "Animals on the Move in the Middle Ages". MAD has organized a series of small conferences connected to animals as material culture over the past fifteen years, starting with a volume edited by Aleks Pluskowski in 2005 entitled *Just Skin and Bones? New Perspectives on Human-Animal Relations in the Historical Past* (BAR International Series). *Medieval Animals on the Move* will be the fifth volume in this series.

MAD is a loose international network of specialists initiated by Alice M. Choyke and Gerhard Jaritz, from Central European University, in 2005. The aim of this database is to encourage scholarly integration of textual, iconographic, and archaeological sources on medieval animals based on proper understanding of how data are generated differently in various disciplines. The hope was to facilitate cross-disciplinary discourse based on properly understood, multi-disciplinary information. The intent of the database is therefore more educational than comprehensive. In addition to the database, there is also a blog (https://mad.hypotheses.org/) and a Facebook page.

MAD's regularly organized academic meetings promote its aims by bringing together researchers from disparate fields dealing with the rich medieval history of animal–human relationships. The proceedings of some previous meetings held in London (2007), Madrid (2013), Vienna (2010), and Budapest (2017) have also been published. Since 2007, MAD has also been represented by its own sessions at the annual meetings of the International Medieval Congress held in Leeds, UK, and is linked with the International Council for Archaeozoology and its numerous working groups.

The goal of the Stockholm meeting was to highlight the historical development and dynamics of the entangled nature of animal–human relationships in medieval times. Movement, as the guiding theme of the conference that ultimately inspired this volume, represents part of the way animals could be understood in the Middle Ages. Medieval people defined beasts by the way they physically moved and behaved in the natural world but also by the way they moved onto the written page and images from ideas circulating about them derived from ancient and contemporaneous authorities. These notions changed depending on the author and the cultural traditions of the part of Europe these manuscripts were produced.

The volume offers a review of Medieval and Early Modern Age cultural attitudes toward animals, reflecting diversity in social life. It is aimed, not only at researchers and students exploring the history of animals, but also at a broader readership interested in how our attitudes toward the animal world have evolved over centuries in a variety of cultural contexts.

The chapters included contribute to integrating three basic branches in medieval studies: archaeology, history (comprising both documentary and literary sources), as well as iconography.

These differing sources have traditionally been studied using different paradigms. The integrated approach in this book is meant to strengthen awareness of the complex interplay between the histories of nature and culture in scholarship. In addition to being multi-disciplinary, the volume is emphatically international, with authors representing research in Austria, China, France, Hungary, Ireland, Italy, Serbia, Sweden, and Switzerland.

Stockholm, Sweden László Bartosiewicz
Vienna, Austria Alice M. Choyke

Contents

1 **Introduction: Animals Stepping off the Page** 1
László Bartosiewicz and Alice M. Choyke

2 **The Forgotten Pigs and Goats of Iceland in a North Atlantic Context** 13
Bernadette McCooey

3 **Imperial Horse Policy and the Publication of Equine Veterinary Medicine Books in Ming China: A Case Study on *Yuanheng Liaomaji*** 41
Zhexin Xu

4 **Medieval Animals: The Fast and the Slow** 67
Gerhard Jaritz

5 **Animals Between Authors and the Natural World in Giovanni da San Gimignano's *Liber de exemplis et similitudinibus rerum*** 81
Beatrice Amelotti

6 **Always Angular and Never Straight: Medieval Snakes in Human Graves?** 95
Monika Milosavljević

7 **Perpetual Preys: Pursuing the Bonnacon Across Space and Time** 121
Zsuzsanna Papp Reed

8 **What's in a Noun? A Short *Caveat* Regarding the Difficulties of Identifying Medieval Animals in Texts** 151
Richard Trachsler

9 **Exotic Encounters: Vikings and Faraway Species in Motion** 167
Csete Katona

10 **The Question of Feathers in the Early Modern Cabinet of Curiosities (Sixteenth–Seventeenth Centuries)** 179
Myriam Marrache-Gouraud

Index 189

List of Contributors

Beatrice Amelotti Dipartimento di Studi Umanistici, Università degli Studi di Pavia, Pavia, Italy

László Bartosiewicz Osteoarchaeological Research Laboratory, Stockholm University, Stockholm, Sweden

Alice M. Choyke Department of Medieval Studies, Central European University, Vienna, Austria

Gerhard Jaritz Department of Medieval Studies, Central European University, Vienna, Austria

Csete Katona Department of Medieval Studies, Central European University, Vienna, Austria

Myriam Marrache-Gouraud Laboratoire Forellis, Université de Poitiers, UFR Lettres et Langues, Département Lettres, Poitiers Cedex 9, France

Bernadette McCooey Independent Scholar, Birmingham, UK

Monika Milosavljević Faculty of Philosophy, Department of Archaeology, University of Belgrade, Belgrade, Serbia

Zsuzsanna Papp Reed Department of Medieval Studies, Central European University, Vienna, Austria

Richard Trachsler Romanisches Seminar, Universität Zürich, Zürich, Schweiz

Zhexin Xu Fachbereich Geschichte, Universität Salzburg, Salzburg, Austria

List of Figures

Fig. 1.1	Michael Bernhard Valentini's illustrated summary of unicorns. Internet Archive, https://archive.org/stream/mobot31753000820560#page/n15/mode/1up	5
Fig. 1.2	The greatest present-day circumpolar distribution of narwhal populations. Striped areas indicate rare occurrences. The North Pole is marked by a cross in the center of the map. Image in the public domain. https://commons.wikimedia.org/wiki/File:Narwhal_distribution_map.png	6
Fig. 1.3	Ram of the Texel breed in England with a single vestigial horn growing in the middle of its forehead. (Photo: Courtesy Louise Gidney)	8
Fig. 1.4	Osteological evidence: present-day sheep skull with the stub of a single rudimentary horn (Osteoarchaeological Research Laboratory, Stockholm University, Inv. No. KO-700) from the Scan Farmek slaughterhouse, Uppsala 1987. Scale: 5 cm	9
Fig. 2.1	The geographical distribution of archaeological and documentary sources in Iceland. Base map attribution: http://www.vidiani.com/, Creative Commons. Locations shown: 1. Hóll, 2. Staður, 3. Holt, 4. Núpur, 5. Mýrar, 6. Hrafneyri, 7. Tálknafjörður, 8. Saurbær, 9. Gufudalur, 10. Reykhólar, 11. Staðarfell, 12. Hólar, 13. Reykir, 14. Árskógur, 15. Oddstaðir, 16. Gásir, 17. Munkþverá, 18. Gnúpufell, 19. Illugastaðir, 20. Háls, 21. Vatnsleysa, 22. Laufás, 23. Hofstaðir, 24. Sveigakot, 25. Svalbarð, 26. Hofteig, 27. Skriða, 28. Þykkvabær, 29. Stóraborg, 30. Bergþórshvoll, 31. Skálholt, 32. Alþing, 33. Bessastaðir, 34. Viðey, 35. Reykholt	18

Fig. 3.1 The study area (highlighted) within Ming Dynasty China. Yellow lines between imperial and provincial capital cities generally illustrate the imperial postal system based on identification of postal stations from around 1580, most of which kept horses. Base map: Wikimedia Commons, image in the public domain after: https://commons.wikimedia.org/wiki/File:%C5%98%C3%AD%C5%A1e_Ming.png 43

Fig. 3.2 The number of imperial horses kept in local pastures within the Nanjing Metropolitan area (*Nan Zhili*) around 1550. (Source: *Nanjing Taipusi zhi*. Original drawing) 54

Fig. 3.3 Postal stations within Nanzhili recorded in *Yitong Lucheng Tuji* (1570). (Source: Huang Bian ed. 1994. Original drawing) 56

Fig. 3.4 The quantity and average area of local pastures in Nanzhili's prefectures under the supervision of *Taipu si* (Court of Imperial Stud) at Nanjing, around 1550. (Source: *Nanjing Taipusi zhi*. Original drawing) 58

Fig. 4.1 The ass as part of the personification of the capital sin of sloth. Wall painting, end of the fourteenth century, parish church St. James, Levoča (Slovakia). (Photo: Institut für Realienkunde, University of Salzburg (Austria)) 73

Fig. 4.2 *Concordantiae caritatis*, Library of the Cistercian house of Lilienfeld (Lower Austria), cod. 151, fol. 32v (detail), the *ludolacra*: *Ad loca longa salis, ludolacra, quatuor alis* (Ludolacra, with four fins, you leap to distant places). (Photo: Institut für Realienkunde, University of Salzburg (Austria)) 75

Fig. 4.3 *Concordantiae caritatis*, Library of the Cistercian house of Lilienfeld (Lower Austria), cod. 151, fol. 203v (detail), the squirrel: *Cauda volat solus fit et agilis asperiolus* (The swift squirrel flies just with its tail). (Photo: Institut für Realienkunde, University of Salzburg (Austria)) 76

Fig. 4.4 *Concordantiae caritatis*, Library of the Cistercian house of Lilienfeld (Lower Austria), cod. 151, fol. 114v (detail), the eel: *Liberat angwillam, si stricta manus tenet illam* (When holding it strongly, the hand frees the eel). (Photo: Institut für Realienkunde, University of Salzburg (Austria)) 77

Fig. 5.1 Dominicans and dogs with their black and white habits depicted by Andrea di Bonaiuto, Cappellone degli Spagnoli, Santa Maria Novella, Florence. Detail. (Wikimedia commons. Image in the public domain). https://commons.wikimedia.org/wiki/File:Andrea_Bonaiuti-The_Church_as_the_Path_to_Salvation_(detail_).jpg 86

Fig. 5.2 Nineteenth century scientific drawing of beaver feet (Boutell 1869, p. 23 No. 40–41. Wikimedia commons. Image in the public domain). https://commons.wikimedia.org/wiki/File:Picture_Natural_History_-_No_40_41_42_-_Beaver_feet_and_tail.png 89
Fig. 5.3 Bear licking her cub into a form. Source: Den Haag, Huis van het boek, 10 B 25, folio 11v. With permission of Huis van het boek 91
Fig. 6.1 The placement of the snake skeletons (graves no. 4, 11, 24, 73, 53, 37) at Ravna–Slog. Visualization by Nikola Stepković 98
Fig. 6.2 Archaeological artifacts associated with snake remains (graves no. 4, 11, 73). Visualization by Nikola Stepković 99
Fig. 7.1 *Bonacon* in a Second Family Bestiary, fifteenth-century English manuscript (Copenhagen, Kongelige Bibliotek, Gl. kgl. S. 1633 4°, Folio 10r. Image in the public domain). http://www.kb.dk/permalink/2006/manus/221/eng/10+recto/?var 129
Fig. 7.2 *Bonacus*. Illumination from *Der naturen bloeme* by Jacob van Maerlant between ca 1340–1350 (The Hague, Koninklijke Bibliotheek, KB KA 16, fol. 47v. Image in the public domain). http://manuscripts.kb.nl/zoom/BYVANCKB%3Amimi_ka16%3A047v_min_b1 133
Fig. 7.3 Bonacon in the *Rochester Bestiary*, British Library, Royal MS 12 F. xiii, fol. 16r. (Image in the public domain). https://commons.wikimedia.org/wiki/File:RochesterBestiary_detail_Bonnacon.jpg 141
Fig. 8.1 Images of lynx in the *Concordantiae caritatis* (1349–1351), Library of the Cistercian house of Lilienfeld (Lower Austria), cod. 151, fol. 31v. Photo: Institut für Realienkunde, University of Salzburg (Austria) 159
Fig. 9.1 Horses unloaded from a longship on the Bayeux tapestry (Source: https://commons.wikimedia.org/wiki/File:BayeuxTapestryScene39.jpg [accessed: 01.02.2018.]) 169
Fig. 9.2 Walrus tusk from Iceland (McGovern 2011; With the author's kind permission) 171
Fig. 9.3 Walrus tusk crosier from the Veszprémvölgy convent in Hungary (Fülöp and Koppány 2004. With the authors' kind permission) 172

Fig. 10.1 Ferrante Imperato's cabinet of curiosities displayed at the Palazzo Gravina, Naples. From *Dell'Historia Naturale* (Imperato 1599; Wikimedia commons. Image in the public domain). https://commons.wikimedia.org/wiki/File:RitrattoMuseoFerranteImperato.jpg 180
Fig. 10.2 Feathered decoration adorning the bodies of Native Americans was treated by Aldrovandi along with exotic birds of the New World (Aldrovandi 1610. L. XI, chap. 1, p. 334; ETH-Bibliothek, Zürich. Image in the public domain). https://www.e-rara.ch/zut/wihibe/content/zoom/4149659 185

List of Tables

Table 2.1	Summary of documentary evidence for the presence of pigs	24
Table 2.2	Summary of documentary evidence for the presence of goats	28
Table 3.1	Structure of *Yuanheng Liaomaji*	47
Table 3.2	Ming governmental institutes responsible for horse administration (horse policy *"ma zheng"* in *Ming shi*, 2270–2277)	50
Table 3.3	Demographic data and the number of registered horses in Nanzhili's prefectures (Source: TPSZ). One *qin* equalled ca 56,655 m² during the Ming era	58
Table 6.1	Graves containing snake skeletons at Ravna–Slog. Ages are given in years	101
Table 7.1	Commonalities and differences between *bonasus* descriptions in three sources	124

CHAPTER 1

Introduction: Animals Stepping off the Page

László Bartosiewicz and Alice M. Choyke

In his twelfth-century poem, *De incarnatione Christi*, the theologian and neo-Platonist, Alain de Lille directed attention to the significance of animals beyond their physical being stating that all the world's creatures, within a book or a picture, mirror us (*Omnis mundi creatura /quasi liber et picture /nobis est in speculum*). In other words, observing animals enhances our understanding of the human condition. The global relevance of animals is also illustrated by the much-discussed statement by Claude Lévi-Strauss (1962), that animal species are chosen as totems not simply as "good to eat" but as "good to think with" (*Les espèces sont choisies non commes bonnes à manger, mais comme bonnes à penser*).

Following initial research into medieval animal remains by zoologists, archaeologists, and historians (e. g. Bökönyi 1995; Bartosiewicz 1995; Pluskowski ed. 2002, 2005, 2007), Harriet Ritvo's article *On the Animal Turn* (2007) finally put the spotlight on animals within general

L. Bartosiewicz (✉)
Osteoarchaeological Research Laboratory, Stockholm University, Stockholm, Sweden
e-mail: laszlo.bartosiewicz@ofl.su.se

A. M. Choyke
Department of Medieval Studies, Central European University, Vienna, Austria
e-mail: choyke@ceu.edu

L. Bartosiewicz, A. M. Choyke (eds.), *Medieval Animals on the Move*, https://doi.org/10.1007/978-3-030-63888-7_1

humanistic studies. As she puts it: “Its very marginality allows the study of animals to challenge settled assumptions and relationships”. Animals play key roles in the understanding of long-term environmental and social processes, representing a vital part of medieval and Early Modern Age cultural heritage in Europe and beyond. However, a subsequent upsurge in relevant research has revealed how indispensable a transdisciplinary approach is in exploiting synergies between the various research areas concerned (Kucera and Kunst eds 2010, García et al., eds 2013; Choyke and Jaritz eds 2017).

Within these thought spheres, ideas about animals continually change their meanings in different contexts. Animals move from book to image and literally transverse the earth through human intentionality, from high mountain pastures to valley settlements, from rural to urban space, from sacred to secular space and from far-away exotics to everyday creatures living close to people. In this change and movement, animals possess a particular agency which in turn influences the way humans have always perceived them as both individuals and as a group in different situations.

Movement, in the broadest sense, was chosen as an organizing principle of this volume. First, the active physical movement of medieval animals is considered, including small stock as well as horses. Archival sources, confirmed by archaeological evidence show how domestic animals accompanied sea-faring peoples to newly conquered territories where theo hostile natural environment was sometimes unsuitable for their long-term survival (McCooey, this volume). Movement by animal herds driven over long distances on land is linked to mobile pastoralism and livestock trade. Across the vast Eurasian steppe belt, horses enabled rulers to transverse and tightly control people living at great distances that would have been impossible on foot. The key strategic importance of this “technical” development is clearly reflected in the contemporaneous veterinary literature from Ming Dynasty China (Xu, this volume).

Transit has primarily been interpreted in its spatial sense—through animals as actors in local and regional transport, by studying travelogues from distant voyages, or tracing the itinerant movements of tinkers and other craftspeople. It is clear that spatial movement must also be understood in terms of diachronic development. The conjunction of distances encompassed by human action intrinsically compresses temporal distance (Helms 1993). Reflecting on the slow or fast nature of various representatives of the animal world played an important role that not only helped medieval people interpret natural phenomena but assisted them to understand themselves (Jaritz, this volume).

Images of animals also leap across wide spaces of the human imagination of the world they lived in on maps of the known world and bestiaries depicting all the known creatures on earth along with their ascribed characteristics. This enabled contemporary cultures to tap into the creative forces achieved by simple, physical movement and look at the consequences of animals being continuously appraised and interpreted in new and multi-faceted cultural contexts.

In the meantime, mundane animals living at arm's length, show extreme "mobility" in a non-spatial, symbolic sense. This is clearly illustrated by the diverse metaphoric roles played by dogs in the global ethnolinguistic record (Bartosiewicz 1998, 73). These roles may range between emotionally charged extremes, frequently rooted in medieval oral tradition. Positive dog imagery occurs in Christian sacral iconography (Bartosiewicz 2011, 222) and is also thoroughly discussed in clerical writings (Amelotti, this volume). Even in the case of this amply documented animal, however, explanation can be found in the archaic-looking medieval custom of burying puppies covered by pots in Hungary (Daróczi-Szabó 2010, 213; an alternative to this apparently belief-driven archaeological phenomenon includes hen skeletons).

The opposite of versatile cognitive transformations of dogs is shown by common, frequently commensal, animals (rodents, amphibians, reptiles) that "creeped" into the cultural sphere but left little trace in the historical record (Nickel 2009, 180). For example, known perceptions of the snake as a healer in the Aesculapian tradition and as a diabolic creature in the medieval Biblical sense are of limited help in their interpretation when encountered in medieval burials with no documented explanation (Miloslavjević, this volume). In the absence of directly relevant written sources, they represent what would be a typical "prehistoric" enigma for archaeologists, hardly ever encountered by historians.

Our present-day perceptions contrasting imaginary and so-called "real" animals may easily become blurred, especially when certain beasts were known only by word of mouth or from stories. For a painter of illuminations hidden away in a monastery, an elephant (native to far-away lands in India and Africa) or a wolf (long extinct in many parts of western and central Europe due to habitat loss and hunting), became more important as moral exemplars since such artisans would practically never have had the chance to see either species during their lifetime. For that matter, the reality of a bonnacon, a bull-like creature who uses its own dung as a defensive weapon (Papp Reed, this volume) and a creature described in travelogues, would have appeared just as likely a beast as a cow from the local village. Only the description and geographical context seem to have differed.

The unicorn, recently resurrected in commercial "My Little Pony" iterations, is perhaps the best-known example of medieval beasts surviving in the popular modern mind. It is a creature that embodies the liminal nature of animals in the human imagination. Following several transformations since Antiquity (Waters 2013) and throughout the Middle Ages, naturalistic late medieval and early modern depictions of this creature usually show a horse-goat like animal with a single horn, mane, and cloven hooves (e.g. Gesner 1565). A unicorn also occurs among other large game animals in the 1542 German edition of the cookbook compiled by Platina Cremonensis (Fahrenkamp 1986, 34). Scientific efforts to identify unicorns culminated in research by Otto von Guericke (1602–1686), the respected scholar, best known for having demonstrated the existence of vacuum using his famous Magdeburg hemispheres in 1650. In 1678, he assembled a skeleton of what he thought was a unicorn, using paleontological finds of various extinct ungulates recovered from a sinkhole near Quedlinburg in Zeunickenberg (Lower Saxony). A sketch of the "*Quedlinburger Einhorn*" was published by Michael Bernhard Valentini (1704, 841) as "*Unicornu fossile*" in the company of "*Unicornu fictitium*" and "*Unicornu marinum*", an actual narwhal (*Monodon monoceros* Linnaeus, 1758; Fig. 1.1). The straight, spirally contoured horn itself, "*Unicornu officinale*" was sought after as a key ingredient for medicine. In his 1539 *Carta Marina*, Olaus Magnus, archbishop of Uppsala (Sweden) actually describes a sea unicorn living in the Iceland and Greenland seas. In his 1555 book, the image of this fish-like creature with a horn protruding from its forehead is already labeled *narval* (Olaus Magnus 1555). (Anatomically, the "horn" is the protruding upper left canine tooth which breaks through the upper lip in the males of this whale species). This important step toward realistic presentation is a reminder of a major problem plaguing the interpretation of medieval texts mentioning animals: we frequently just do not know what modern animal the texts actually refer to. The challenge is apparent not only in the case of imaginary creatures, but also for numerous animals that should be perfectly identifiable by conventional Linnaean taxonomic standards (Traschler, this volume).

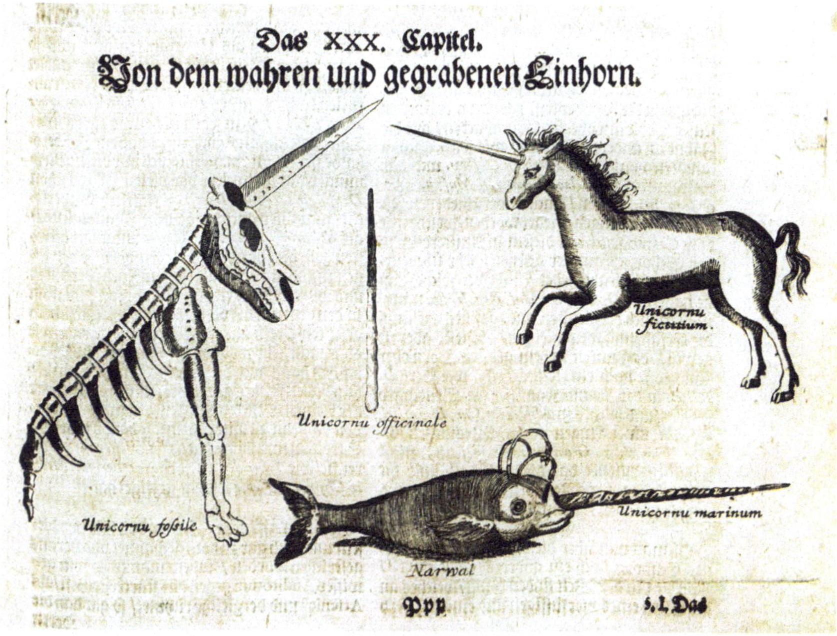

Fig. 1.1 Michael Bernhard Valentini's illustrated summary of unicorns. Internet Archive, https://archive.org/stream/mobot31753000820560#page/n15/mode/1up

Another dimension of animal transitions in a physical sense seems more passive as it comprises the movement of everyday animal products and body parts such as exotic trophies. This latter aspect of medieval animals poses questions regarding the perception of animals and leading to the circulation and valorization of certain materials originating from animals never encountered in everyday life.

The Atlantic walrus (*Odobenus rosmarus* Linnaeus, 1758) would have been visible to anyone who had at least visited northern Norway: during the Norse Period, it was hunted in Greenland, Iceland and the Barents Sea located between Norway and Russia (Smirnova 2001, 9; Pierce 2009, 56; McCooey, this volume; Katona, this volume). Elusive narwhals, however, were available mostly in the Arctic around Greenland and toward Canada as well as near islands off the shores of Siberia. Their proximity to Europe seems to be the occasional extension of the latter area of eastern distribution: it

Fig. 1.2 The greatest present-day circumpolar distribution of narwhal populations. Striped areas indicate rare occurrences. The North Pole is marked by a cross in the center of the map. Image in the public domain. https://commons.wikimedia.org/wiki/File:Narwhal_distribution_map.png

was limited to seasonal occurrences in the Barents Sea (Fig. 1.2). The rare sightings of narwhals by Europeans thus, probably contributed to the myth of the unicorn. Some hypothesize that putative Viking whalers harvested the best part of their catch to be sold as unicorn horns (Humphreys 1953, 17; Christen and Christen 2011, 137) or that Greenland Norse traders forwarded them from the local Inuits further south (Pluskowski 2004, 297–298).

However, evidence is yet to be found of both early whaling and Inuit-Norse trade contacts. On the other hand, such tusks could also be scavenged from beached narwhals (Gaborit-Chopin 1978, 54). Dectot (2018, 170) finds it most plausible that the Norse in Greenland hit upon the remains of beached narwhals' heads and tusks forced to the shore by killer whales, without the live narwhal having actually been sighted. Those tusks could have been gathered and exported, the Norse themselves believing that they had discovered dead unicorns.

Whatever its origins may have been, a lucrative trade in narwhal ivory emerged from conflating conquest and folklore with commercial interests and scientific endeavor (Shepard 1930; Beer 1972). Objects made from this exotic raw material eventually arrived in royal courts, various church treasuries and later, during the Renaissance, in secular cabinets of curiosities designated to display a wide range of exotic animal remains as tokens of knowledge and high-status self-representation (Marrache-Gouraud, this volume).

Already by the ninth century, it is evident that narwhal tusks "formed an important article of trade in the northeast of Europe, that they were known as fish-teeth, and that they were traded to the Turks, and probably reached also inner Asia during the Middle Ages" (Laufer 1913, 338). In fact, there are only a few years between the first mentions of a mysterious animal, *guduxi* (Chinese sources) and *khutū* (Islamic sources), possibly referring to narwhal, both dating around the mid-tenth century (King 2013, 263). This most sought-after part of the original animal, however, moved not only physically over space; it also transited conceptually from scavenged or hunted marine game to miraculous evidence of the unicorn's material existence.

As Aleks Pluskowski (2004, 293) pointed out,

> It is useful to identify an artefact described as a unicorn horn as a narwhal tusk in that it provides information about its geographical and cultural origin, however it is equally useful and essential to consider its identification as a unicorn—touching not only on questions of morphology… but also on conceptualizations in different geographical and cultural contexts. In this respect a zooarchaeological approach treats the physical remains of exotica … as material culture, and more specifically as particular kinds of artefacts shaped by, but also shaping human responses.

Meanwhile, a completely different dimension of zoological reality may also have lain behind encounters with live unicorns. Although not mistaken for the precious raw material originating from narwhals, inherited

Fig. 1.3 Ram of the Texel breed in England with a single vestigial horn growing in the middle of its forehead. (Photo: Courtesy Louise Gidney)

anomalies of horn conformation occur in common domesticates: sheep, goat, and sometimes cattle. Mutations manifested in curious features such as the rudimentary single horn stub in Texel sheep shown in Fig. 1.3 could occur any time. It is but a short step to imagine that knowledge of such rarely occurring mutations in domestic animals became easily entangled with notions of animals known through texts and images, reinforcing ideas and coeval interpretations of actual zoological phenomena. Polyceraty in medieval sheep has been documented from numerous archaeological sites (e.g. Putelat 2005). Hacking marks on the forehead of a medieval four-horned ram skull recovered in the royal capital of Buda in Hungary suggest that an effort was made to transform the horns into a trophy (Daróczi-Szabó and Daróczi-Szabó 2018, 249, Fig. 16.3). Given such genetic anomalies, single-horned individuals may also have occurred in the same way as they do in present-day sheep (Fig. 1.4). Georges-Louis

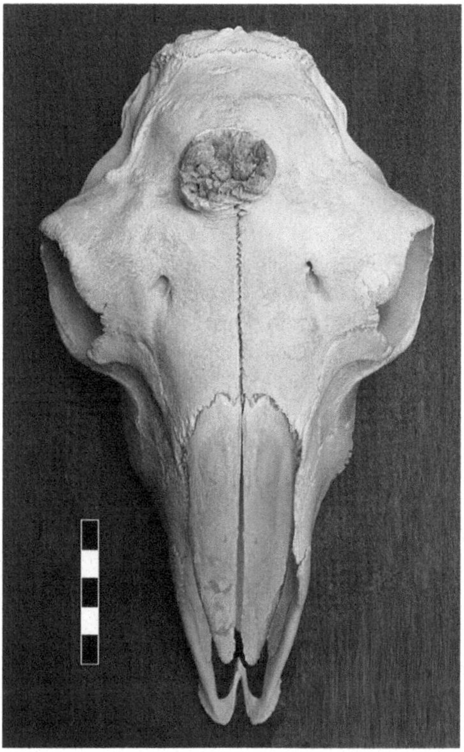

Fig. 1.4 Osteological evidence: present-day sheep skull with the stub of a single rudimentary horn (Osteoarchaeological Research Laboratory, Stockholm University, Inv. No. KO-700) from the Scan Farmek slaughterhouse, Uppsala 1987. Scale: 5 cm

Leclerc, the earl of Buffon (1755, 65), published the image of a three-horned ram from Iceland noting that such beasts were immediately exported to Copenhagen to be sold at a considerable price.

Starting in Viking Period Iceland, our book returns to the same region in the time of Enlightenment, illustrating that animals have always been important in culture, in both their materiality and their varied symbolic roles. This volume offers a review of some examples illustrating medieval attitudes toward animals between the wake of Late Antiquity and the emergence of the Renaissance. Our aim is to present integrated research on written sources, archaeological animal bone finds, and iconographic data. By maintaining a vivid transdisciplinary dialogue, medievalists and zoologists from very different academic backgrounds can expose new aspects of the way people perceived, treated, and used animals in the Middle Ages. The nature of the resulting complex and tangled interpretations is dependent on shifting definitions of what a particular animal means to both present-day researchers and in the minds of people from different social and geographical backgrounds in the Middle Ages.

References

Bartosiewicz, László. 1995. *Animals in the Urban Landscape in the Wake of the Middle Ages*. Oxford: British Archaeological Reports, International Series 609.

Bartosiewicz, László. "Attitudes to pets in the ethnolinguistic record." In *Man and the Animal World. Studies in memoriam Sándor Bökönyi*, edited by Peter Anreiter, László Bartosiewicz, Erzsébet Jerem, and Wolfgang Meid, 65–78. Budapest: Archaeolingua Kiadó, 1998.

Bartosiewicz, László. "'Stone Dead': Dogs in a Medieval Sacral Space." In *The Ritual Killing and Burial of Animals: European Perspectives*, edited by Aleks Pluskowski, 220–229. Oxford: Oxbow Books, 2011.

Beer, Rüdiger Robert. 1972. *Einhorn: Fabelwelt und Wirklichkeit*. München: Georg D. W. Callwey.

Bökönyi, Sándor. "The Development of Stockbreeding and Herding in Medieval Europe." In *Agriculture in the Middle Ages: Technology, Practice, and Representation*, edited by Dale Sweeney, 41–61. Philadelphia, PA: University of Pennsylvania Press, 1995.

Buffon, Comte de, 1755. *Histoire Naturelle, générale et particulière, avec la description du Cabinet du Roi*. Vol. V. Paris: L'Imprimerie Royale.

Choyke, Alice M. and Jaritz, Gerhard eds. 2017. *Animaltown: Beasts in Medieval Urban Space*. British Archaeological Reports, International Series 2858. Oxford: Archaeopress.

Christen, Ardem G., and Christen, Joan A. "The unicorn and the narwhal: a tale of the tooth." *Journal of the History of Dentistry* Vol. 59, no. 3 (2011). 135–142.

Daróczi-Szabó, Márta. "Pets in pots: Superstitious belief in a medieval Christian (12th–14th c.) village in Hungary." In *Anthropological Approaches to Zooarchaeology: Colonialism, Complexity and Animal Transformations*, edited by Douglas V. Campana, Pamela Crabtree, Susan D. deFrance, Justin Lev-Tov, Alice M. Choyke, 211–215. Oxford: Oxbow, 2010.

Daróczi-Szabó, Márta and Daróczi-Szabó, László. "Medieval Multi-Horned Sheep from Present-Day Budapest, Hungary." In *Care or Neglect? Evidence of Animal Disease in Archaeology*, edited by László Bartosiewicz, and Erika Gál, 247–255. Oxford & Philadelphia: Oxbow, 2018.

Dectot, Xavier. "When ivory came from the seas. On some traits of the trade of raw and carved sea-mammal ivories in the Middle Ages.". In *Animaux aquatiques et monstres des mers septentrionales (imaginer, connaître, exploiter, de l'Antiquité à 1600)*, edited by Catherine Jacquemard, Brigitte Gauvin, Marie-Agnès Lucas-Avenel, Benoît Clavel, and Thierry Buquet. *Anthropozoologica* Vol. 53, no. 14 (2018). 159–174.

Fahrenkamp, Hans J. 1986. *Wie man eyn teutsches Mannsbild bey Kräfften hält.* München: Orbis Verlag.

Gaborit-Chopin, Danielle. 1978. *Ivoires du Moyen Âge*. Fribourg: Office du Livre.

García, Francisco de A., Mónica A. Walker Vadillo, and María V. Chico Picaza eds. 2013. *Animals and Otherness in the Middle Ages.* British Archaeological Reports, International Series 2500. Oxford: Archaeopress.

Gesner, Conrad. 1565. *De omni rerum fossilium genere, gemmis, lapidibus metallis, et huiusmodi, libri aliquot, plerique nunc primum editi.* Tiguri: Christophorus Froschoverus.

Helms, Mary W. 1993. *Craft and the Kingly Ideal. Art, Trade, and Power.* Austin TX: University of Texas Press.

Humphreys, Humphrey. "The horn of the unicorn." *Antiquity* Vol. 27, no. 105 (1953). 15–19.

King, Anya. "Early Islamic sources on the Kitan Liao: the role of trade." *Journal of Song-Yuan Studies* Vol. 43 (2013). 253–271.

Kucera, Matthias and Kunst, Günther-Karl eds. 2010. *Bestial Mirrors. Using animals to construct human identities in Medieval Europe.* ViaVIAS 2010/3. Wien: Vienna Institute for Archaeological Science.

Laufer, Berthold. "Arabic and Chinese Trade in Walrus and Narwhal Ivory" *T'oung Pao* Vol. 14, no. 3 (1913). 315–370.

Lévi-Strauss, Claude. 1962. *Le Totémisme aujourd'hui.* Mythes et religions 42. Paris: Presses universitaires de France.

Nickel, Réka. "A 'bűbájos asszonyállat'. Állattá változások a Magyarországon lefolytatott kora újkori boszorkányperekben (The charmed 'animal woman'. Animal transformations in Early Modern Age witch-trials in Hungary)." In *Csontvázak a szekrényből. Válogatott tanulmányok a Magyar Archaeozoológusok Visegrádi Találkozóinak anyagából 2002–2009*, edited by László Bartosiewicz, Erika Gál, and István Kováts, 173–186. Budapest: Martin Opitz Kiadó, 2009.

Olaus Magnus. 1555. *Historia de Gentibus Septentrionalibus.* Romae.

Pierce, Elizabeth. "Walrus hunting and the ivory trade in Early Iceland." *Archaeologia Islandica* Vol. 7 (2009). 55–63.

Pluskowski, Aleks ed. 2002. *Medieval Animals.* Vol. 18. Cambridge: Archaeological Review from Cambridge.

Pluskowski, Aleks. "Narwhals or unicorns? Exotic animals as material culture in medieval Europe." *European Journal of Archaeology* Vol. 7, no. 3 (2004). 291–313.

Pluskowski, Aleks. 2005. *Just Skin and Bones? New Perspectives on Human-Animal Relations in the Historical Past.* British Archaeological Reports, International Series 1410. Oxford: Archaeopress.

Pluskowski, Aleks ed. 2007. *Breaking and Shaping Beastly Bodies: Animals as Material Culture in the Middle Ages*. Oxford: Oxbow Books.

Putelat, Olivier. "Le bestiaire polycère." *Revue de Paléobiologie* Vol. spéc. 10 (2005). 293–301.

Ritvo, Harriet. "On the animal turn." *Daedalus* Vol. 136, no. 4 (2007). 121.

Shepard, Odell. 1930. *The Lore of the Unicorn*. London: George Allen & Unwin.

Smirnova, Lyuba. "Utilization of rare bone materials in medieval Novgorod." In *Crafting Bone: Skeletal Technologies through Time and Space*, edited by Alice M. Choyke, and László Bartosiewicz, 9–17. British Archaeological Reports International Series 937. Oxford: Archaeopress, 2001.

Valentini, D. Michael Bernhard. 1704. *Natur und Materialen Kammer auch Oost-Indianische, Send-Schreiben und Rapporten*. Buch III. Franckfurt am Mayn: Johann David Zunner.

Waters, Elyse. "Zoological Analysis of the Unicorn as Described by Classical Authors." *Archeometriai Műhely* Vol. X, no. 3 (2013). 231–236.

CHAPTER 2

The Forgotten Pigs and Goats of Iceland in a North Atlantic Context

Bernadette McCooey

Introduction

When Iceland was settled in the ninth and tenth centuries, a range of domesticated animals were introduced, including cattle, sheep, horses, pigs, goats, poultry, cats, and dogs. By the early eighteenth century, Icelandic farming was generally dominated by sheep, and to a lesser extent cattle, with horses being utilised for their non-edible products. Less is known and discussed about other species of livestock. Within a few centuries of settlement, cattle and sheep had become the most numerous species and discussions about Icelandic farming have been dominated by those two species. Goats and pigs, while initially common, were not numerous and decreased in number, or ceased to exist in the case of pigs. This article aims to tell the story, not about the winners such as sheep and cattle, but about the species that lost out in the long-term adaptation of pastoral farming in Iceland. A fuller understanding of Icelandic agriculture can only be achieved by considering the species that became scarce, not just the species that prospered. To avoid short-sighted explanations of

B. McCooey (✉)
Independent Scholar, Birmingham, UK

L. Bartosiewicz, A. M. Choyke (eds.), *Medieval Animals on the Move*, https://doi.org/10.1007/978-3-030-63888-7_2

processes (Dugmore et al. 2013, 435–436), the chronology of this chapter begins with the settlement and extends to the eighteenth century, when the first countrywide livestock population record was compiled and Iceland began to move into the early modern period.

During the settlement of Iceland, a process of trial and error was undertaken as the settlers adapted to their new land, and this can be seen in the reliance on wild resources such as fish and birds while domestic herds were established (McGovern et al. 2006, 191). In the centuries that followed, Icelanders continued to adapt to their changing environment and climate, utilising both wild and domestic resources. Among their domestic livestock, they altered the numbers and proportions of the species to best respond to their conditions. Any investigation into farming practices must be multi-disciplinary, and as such, a variety of evidence is drawn upon from written documents to animal bones. A brief overview of sources is required before moving on to the discussion.

Material and Methods

Sagas

The most well known of the Icelandic evidence is the sagas, tales of feud and interactions among the elites. The sagas have been divided by modern scholars into several categories, of which two will be considered in this chapter: the Sagas of Icelanders, *Íslendingasögur*, and the Saga of the Sturlungs, *Sturlunga saga*. Both categories were written down in the thirteenth and fourteenth centuries with some surviving in later manuscripts, yet the two categories depict different times. The Sagas of Icelanders portray the tenth and eleventh centuries and a stable system of farming. The Saga of the Sturlungs records events from the twelfth and thirteenth centuries, an increasingly turbulent time in Icelandic society that ultimately led to Iceland's submission to the Norwegian Crown in the late thirteenth century. While differing in focus, these sagas provide everyday details about farming and are used to contextualise other sources. The sagas construct a sense of realism by their level of detail which adds flesh to the bones of the legal texts.

Legal Texts

Four legal texts cover the time period under consideration in this study. The oldest is *Grágás* (Dennis et al. 2000) said to have been created at the

Alþingi, the General Assembly in Iceland, and put into writing in the early twelfth century. As the manuscripts of *Grágás* were private collections of the laws, they differ in the sections recorded but demonstrate a clear interest in prescribing resolutions for all aspects of societal dispute. *Grágas* was once thought to be an indigenous creation, though it is now recognised to have been influenced by twelfth- and thirteenth-century European laws (Sigurðsson 1999, 19). Nevertheless, a comparable amount of information relating to farming is scarcely found anywhere else in sources from the Norse regions. The next legal text is the short-lived *Járnsíða*, introduced when Iceland submitted to the Norwegian Crown. It proved unpopular and was replaced by *Jónsbók* ten years later (Schulman 2010). As *Járnsíða* was close to Norwegian laws and was only briefly in use, it will not be considered here. *Jónsbók* was introduced in 1281 and adhered to until the eighteenth century, though certain clauses were amended several times. *Jónsbók* was closer in content to *Grágás* than *Járnsíða* had been (Schulman 2010, xiv–xv). The final legal text is *Búalög*, known as a household or agricultural law text, because of its focus on household and farming matters. *Búalög* survives in manuscripts dating from the fifteenth century and later. It was used to supplement *Jónsbók* but does not provide information on the keeping of goats and pigs.

Máldagar

The *máldagar* are church-charters recording the property of individual churches. The charters have been dated between the twelfth and sixteenth centuries. Several hundred such documents survive, with most churches only having one or two surviving charters. Occasionally, a church has more charters and where this happens the charters can be dated several centuries apart. The information about livestock contained in these charter sources can include the sex and ages of livestock, permitting a detailed picture to be constructed of the herds. While this information survives for church-owned farms we, unfortunately, do not have comparative records for secular farms.

Documents

Other written documents also survive from the twelfth century onward such as agreements, price-lists, and letters of sale. They provide snap-shots of the people involved and the exchanges that were concluded at specific locations as well as the goods, including livestock that were of sufficient value to be recorded. From these documents we can infer the status of the people and the availability of livestock at a particular farm, which gives an insight into livestock on Icelandic farms in general. For the purpose of this chapter, only those documents concerning pigs and goats around Iceland are included.

Icelandic Scholarship

From the sixteenth to eighteenth centuries we have observational writings by Icelanders, including bishops, scholars, and travellers, that can be described as scholarly in nature. These accounts record various aspects of Icelandic life, history, and geography based on the authors' own experiences and from information gathered from others. Iceland had not yet begun to industrialise and so these writers had exposure to farming activities even if they might not have had hands-on experience.

Due to the rural, pastoral nature of Icelandic society, these writings describing the people, customs, and landscape inevitably mentioned livestock and indicated potential explanations for the rearing of livestock and possible difficulties involved. We therefore have an Icelandic perspective on farming and can be certain that the information is, on the whole, accurate or at the very least reflective of generally held beliefs.

The first attempts to modernise Iceland were undertaken from the mid-eighteenth century onward, but initially proved unsuccessful. As such, the distinction between late medieval and early modern is usually placed between the eighteenth and nineteenth centuries when Reykjavík officially became a town by royal decree and industries, such as fishing and weaving, became full-time occupations for people (Karlsson 2003, 173–176, 182–185).

Jarðabók

While the late seventeenth century saw the compilation of several population registers in Iceland, the *Jarðabók* of Árni Magnússon and Páll Vídalín undertaken between 1702 and 1712 was the first to record both people and livestock on a countrywide basis (Magnússon and Vídalín 1980–1988). There are issues with the register, for example, a smallpox outbreak disrupted the gathering of information between 1707 and 1709. It is thought that this outbreak killed a large proportion of the population and would thus, have affected livestock numbers as well as any comparisons drawn between regions recorded before and after the outbreak. The records for two regions in eastern Iceland have been lost, narrowing the scope of any conclusions with regard to those regions. Finally, there have been questions about the accuracy of the information, as it was compiled for tax purposes and so may have resulted in the under-recording of livestock and property to avoid paying higher taxes (Simpson et al. 2001, 184; Edvardsson 2004, 190). Nevertheless, *Jarðabók* is a goldmine of information about livestock species, herd sizes and composition, and allows an insight for the first time, into livestock on farms across the social spectrum.

Archaeology

The final source of information drawn upon in this article is zooarchaeology, specifically the animal bones excavated from farm middens. These bone assemblages offer evidence of the species present and, to some extent, their relative proportions (see Reitz and Wing 2008 for further explanations about zooarchaeology). An important issue when examining the archaeological evidence for goats is the difficulty of distinguishing between the skeletal remains of goats and sheep. While some remains are clearly distinguishable as either sheep or goat, most are not and the uncertain bones are grouped in the 'caprine' category. In some cases, the combined 'caprine' group can be quite large, though it has recently been cautioned that when no identifiable goat-bones can be clearly distinguished in a collection the presence of goat-bones is unlikely (Baldursdóttir et al. 2017, 3). There are similar issues for distinguishing some pig bones, such as ribs and vertebrae, from similar bones of other comparably sized mammals (Arge et al. 2009, 22). New methods for species identification are being developed, for example, aDNA analysis, but morphological identification is the simplest and most commonly employed method.

Another issue with archaeological excavations is that they have tended to focus on higher status sites from the earlier centuries of Icelandic history. Traditionally, excavations were undertaken at supposed saga sites in order to flesh out the details mentioned in the sagas, with discoveries being explained by events recorded in the sagas. More recently, there has been an interest in the settlement of Iceland and the establishment of a 'new' society, which resulted in a research focus on the wider landscape around settlers' farms, such as the tenth century Hofstaðir in Mývatnssveit (see Fig. 2.1 for site locations). This skewed approach to earlier sites is

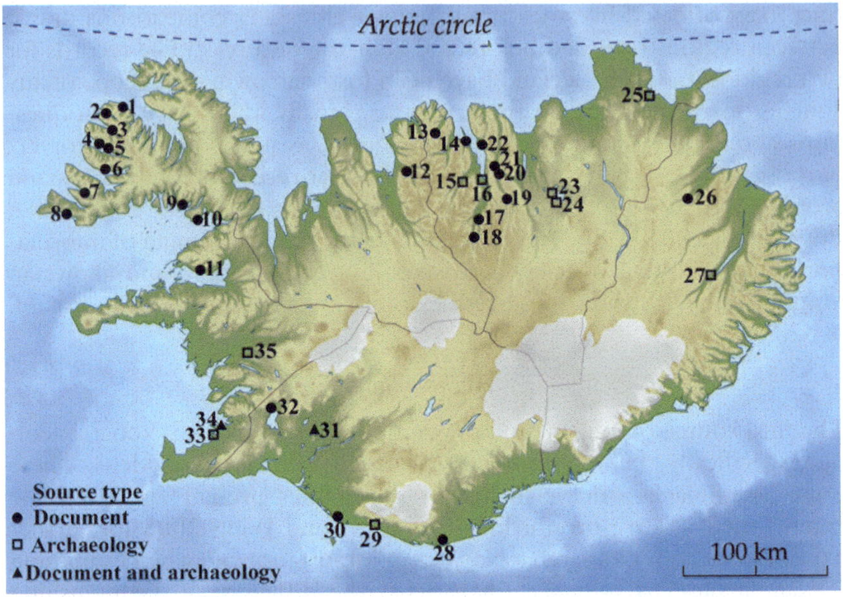

Fig. 2.1 The geographical distribution of archaeological and documentary sources in Iceland. Base map attribution: http://www.vidiani.com/, Creative Commons. Locations shown: 1. Hóll, 2. Staður, 3. Holt, 4. Núpur, 5. Mýrar, 6. Hrafneyri, 7. Tálknafjörður, 8. Saurbær, 9. Gufudalur, 10. Reykhólar, 11. Staðarfell, 12. Hólar, 13. Reykir, 14. Árskógur, 15. Oddstaðir, 16. Gásir, 17. Munkþverá, 18. Gnúpufell, 19. Illugastaðir, 20. Háls, 21. Vatnsleysa, 22. Laufás, 23. Hofstaðir, 24. Sveigakot, 25. Svalbarð, 26. Hofteig, 27. Skriða, 28. Þykkvabær, 29. Stóraborg, 30. Bergþórshvoll, 31. Skálholt, 32. Alþing, 33. Bessastaðir, 34. Viðey, 35. Reykholt

now being corrected with archaeological projects such as Gásir and its hinterlands on the coast of Eyjafjörður, one of the longest fjords in Iceland (Harrison 2013, 2014; Harrison et al. 2008). At present, there are fewer zooarchaeological collections available for the later centuries discussed in this chapter. Moreover, the stability of Icelandic farm locations from the eleventh century onward also hinders investigations because modern farms are usually situated upon older farm remains. The abandonment of modern farms as the Icelandic population becomes more urban may serve to remove this obstacle in the future, especially in remote areas of the country where abandoned farms are becoming numerous.

The written sources and the zooarchaeological evidence represent two ends of the same continuum: production and consumption. The written sources record or indicate the partial or total number of livestock on a farm, the animals that are being reared or owned by the farm. We can make inferences about unrecorded livestock too; for example, if milking goats are recorded, then it can be assumed that there were kids even if they might not be listed in the sources. Whether these kids were slaughtered soon after birth to preserve the milk for human consumption or reared to adulthood is a different matter. The zooarchaeological sources, however, reveal the age animals were slaughtered. The physical evidence also reveals the diets of Icelanders and that they relied on wild resources such as fish, demonstrating that they were not dependent on only one source of food (McGovern et al. 2001, 159).

Because the written sources and zooarchaeological finds show the two ends of the animal husbandry process, they are not directly comparable. The ratio of animal bones is not the same as the ratio of livestock on a farm. Moreover, the timeframes of the written and physical sources are different. A document is written at one point in time, and though it can be later edited, it is still intended to represent that time and place. On the other hand, a midden can be formed over years or decades and represents a much longer process, as reflected in the wide chronological dates assigned to midden contexts. One context in a midden can contain evidence for several changes in farming practices, while a written document is a snapshot of the livestock preferences for a farm at one place in time with little indication of changes in husbandry routines. With an awareness of the issues, some generally applicable to all studies of medieval animal husbandry and others specific to the Icelandic context, we can move on to discuss when and where pigs and goats have been found in Iceland.

Results

According to the legal texts *Grágás* and *Jónsbók*, pigs and goats had contrasting values. Sows, female pigs, were highly valued. A two-winter-old sow (all animals in Iceland being aged by winters rather than years), with nine piglets was valued at one *kúgildi*, the standard abstract unit of livestock value in Iceland. A *kúgildi*, the price of a milking cow, is defined in *Grágás* as 'a cow three years and older, ten-years-old or younger, capable of bearing calves, in milk, horned and free of defects, no worse than an average beast, fit enough to be driven from one district to another at the moving days and giving enough for a calf at milking' (Dennis et al. 2000, K.246, 208) or in the later *Jónsbók*, 'a cow eight-winters old but not younger than two-winters old, healthy and in milk, which has calved during the winter after the feast of the Conversion of St Paul (25th January) no worse than an average cow, fit enough to be driven from one district to another on the moving days' (Schulman 2010, VIII 6, 303).

The two-winter sow with nine piglets held her value throughout the legal texts and the *Alþing* price-lists of c.1100 and c.1280 (Dennis et al. 2000, K.246, 209; Schulman 2010, VIII 6, 305; Jónsson et al. eds 1857–1950: 1, 166; 2, 170). Only one local price-list survives for the whole of Iceland, and that is for the Árnes district in south-west Iceland. This price-list records neither pigs nor goats and omits several other livestock categories, hinting at the individual nature of local price-lists. Unfortunately, we are unable to compare with other districts, but it does show the individual nature that price-lists could have had when compared with the *Alþing* price-lists. The absence of pigs and goats in the Árnes price-list suggests that these species were absent on farms in the Árnes district or were not exchanged regularly enough to be valued. The prescriptive nature of the valuations implies that if the animals were exchanged they would be valued.

In comparison to the value of a sow, *Grágás* assessed goats as:

> 'Six goats with kids and in the same condition as prescribed for ewes equal a cow, as do eight barren goats of three years or older. Eight yearling goats able to feed their kids equal a cow. Eight two-year billy-goats equal a cow, four of them uncastrated and four fully castrated, and six three-year billy-goats equal a cow, half of them fully castrated and half of them uncastrated. A billy-goat of four years and another of two years are worth two nanny-goats, a two-year billy-goat one (nanny-goat). If billy-goats are older than those so far listed, then their value is subject to assessment. Two year-old goats equal one nanny-goat, one a young female or a fully castrated male, the other an uncastrated male or billy-goat for stud.' (Dennis et al. 2000, K.246, 208–209)

These valuations are very close to the valuation listed for sheep, which *Grágás* gives as:

'Six ewes, two of two winters and four older, able to feed their lambs, not losing their wool and with fleece and lambs, equal one cow. Eight ewes completely barren, of three winters or older, equal one cow. Eight two-winter old wethers equal a cow and so do eight yearling ewes able to feed their lambs. Six three-winter wethers equal a cow. A four-winter wether and another of two winters equal two ewes. A two-winter ram is worth one ewe. Twelve winter-old sheep equal one cow. All these animals are to be sound and with fleeces. A ram of three winters and older and a leader-wether, their value is subject to assessment.' (Dennis et al. 2000, K.246, 208)

As can be seen, goats and sheep were assigned similar valuations, except for comments about wool, and so demonstrate that there was little perceived value difference between the species in the centuries after the settlement. Comparable assessments are also found in *Jónsbók* and the *Alþing* price-lists, emphasising that there was nonly a small legal difference in values between sheep and goats that continued up until the eighteenth century. Indeed, an amendment to *Jónsbók* in 1294 states that an owner is to pay damages 'if a ram comes among sheep or a buck among goats' showing that what applied for one species applied for the other (Schulman 2010, *réttarbók* clause 22 1294, 403). While it could be argued that this was a Norwegian-inspired clause introduced into Icelandic law, the introduction indicates that the clause was necessary and goats were still to be found in Iceland, and held similar economic positions according to these valuations.

Returning to pigs, zooarchaeological evidence shows that they were present from the settlement of Iceland in the ninth and tenth centuries, but not abundant. Overall, the general trend was that pigs were no longer present on 'normal' farms, inferred to be non-elite farms, from the mid-tenth century (McGovern et al. 2001, 157) and became rare on all farms after the fourteenth century (Harrison et al. 2008, 109–110). There is evidence for pigs between the sixteenth and eighteenth centuries at some sites, for example Bessastaðir, Viðey, Stóraborg, and Skálholt. These sites range in purpose but are classed as wealthier sites, indeed Skálholt is the bishop's seat for southern Iceland. The late medieval/early modern evidence of pigs at these sites sometimes rests on the recovery of a small number of pig bones: seven pig bones out of a domestic species total of 1784 bones at Bessastaðir (number of identifiable specimens, NISP = 11,179), three out of 594 domestic animal bones at Viðey (NISP = 3218), 11 from a total 2113 domestic animal bones at Stóraborg (NISP = 14,298), and two pig bones in 2151 domestic animal bones at Skálholt (NISP = 5247; Hambrecht

2009, 20). These small quantities offer no firm evidence for the rearing of pigs but instead imply the occasional consumption of pork brought to the site. At Videy, however, the recovery of neo-natal and adult pig bones is an indicator of pig rearing after pigs had become rare in the zooarchaeological record (Amorosi 1996, 411).

As stated previously, the wide chronological intervals represented by deposits make it difficult to be certain when the pig remains were deposited. In addition, the recovery of pig bones from early modern urban sites has been explained by merchants bringing pork or even live pigs to Iceland for fresh meat, as testified in the writings of Eggert Ólafsson and Bjarni Pálsson in the mid-eighteenth century (Ólafsson and Pálsson 1805, 29). Evidence from trading sites may represent imported pork and pigs, and thus, may not be reflective of animal husbandry on Icelandic farms. Some trading and rural sites may have been inter-connected, as it is argued for Gásir, Eyjafjörður, where nearby farms reared livestock to supply Gásir with meat. For example, up until the fourteenth century, cattle appear to have been reared for prime beef at Oddstaðir, Hörgárdalur, for Gásir some 20 km away (Harrison 2014, 130–131). It may have been during trading that Icelandic products were exchanged for foreign cuts of pork, which were then consumed on Icelandic farms (Harrison et al. 2008, 109). This practice may have occurred elsewhere in Iceland and could explain the small numbers of pig bones recovered, as pigs and pork were imported. The problematic nature of pig rearing in Iceland, such as their foraging habits, a lack of suitable pasturage, and their need for housing during winter, would lend support to this assertion. The written sources, however, show that native pig rearing was still undertaken in Iceland.

Perceptions of animals can assist our understanding of their husbandry, and the descriptive sagas show a shift in attitudes towards pig keeping. For example, the infrequent references to pigs in the Sagas of Icelanders illustrate how pigs were seen as beneficial during the first settlement phase when their foraging behaviour initially assisted land clearance. The descriptions of escaped pigs rapidly multiplying in the lush new land reveal how bountiful Iceland was seen as but also how pigs were perceived to be a quickly reproducing source of meat (Benediktsson ed. 1986, chapter 116, 158 and chapter 179, 220; Sveinsson ed. 1939, VIII chapter 15, 43). Rapidly breeding pigs could provide Icelanders with meat while they focused on establishing themselves and their herds of slower breeding cattle and sheep. Saga stories of pigs set in the centuries after settlement, however, reveal a change in attitudes towards pigs. Gone are the advantages as pigs became problematic and in some cases were seen as potentially dangerous supernatural creatures. In one of the stories in the Saga of the Sturlungs set in the mid-twelfth century, a sow escapes from the sty and breaks into a farmhouse where she fatally bites a small child before returning to her sty (Jóhannesson et al. eds 1946 I, chapter 4, 123). A change has taken place from pigs being allowed to graze to an emphasis that pigs must be securely housed. Further support for this point is found in *Grágás*, which states that pigs were not permitted on communal land, and a pig without a nose-ring could be killed if found trespassing (Dennis et al. 2000, K.202, 133 and K.207, 139).

Later, the church-charters and *Jarðabók* provide geographical evidence of pigs, or rather the exclusion of pigs from certain grazing land, but do not record the presence of pigs (see Table 2.1). We can therefore gather that pigs had been present on these farms at some time in the past, but by the time of the documents pigs had ceased to be kept. The church-charters and *Jarðabók* copied the earlier documents word-for-word to preserve the rights of these farms to those grazing areas and so fossilised the old wordings as proof of the antiquity of ownership.

Some documents do record the continued existence of pigs in Iceland. Probably the most famous of these documents is the confiscation record for the wealthy Guðmundur Arason in 1446. The document is a valuable source because it lists all livestock owned on his six farms in the Westfjörðs. At Núpur ten old pigs and two sows with seven piglets each were recorded, while Saubær had nine old pigs and two sows with piglets, and Reykhólar had eight old sows and piglets (Jónsson et al. eds 1857–1950, 4, 684, 687 and 869). Though it is uncertain what was meant by old, it is clear from the presence of piglets that pigs were still being bred, and that sows and boars must both have been present. In the valuation sections of *Grágás* and *Jónsbók* mentioned previously, a sow of two-winters or older and nine piglets were valued the same as a milking cow. This valuation demonstrates the value of breeding pigs but also that litters could be as large as nine piglets, giving the potential for Guðmundur Arason to have owned up to a hundred piglets (eight sows with unspecified numbers of piglets plus the 14 piglets at Núpur) in addition to his 31 adult pigs, which would have been a viable breeding population of pigs in the Westfjörðs of Iceland in the mid-fifteenth century.

The church-charters provide another avenue for investigating pigs in Iceland, and Table 2.1 summarises the numbers and locations of pig keeping and grazing.

Table 2.1 Summary of documentary evidence for the presence of pigs

Source	*Year*	*Location*	*Description*
DI 4, 7	1300	Kvigindafells and Laugardalur, Tálknafjörður, Westfjörðs	horse and pigs not allowed on common pasture except when the other farm permits
DI 2, 575	1324	Staður, Súgandafjörður	sow with nine piglets
DI 2, 617	1327	Hóll, Bolungarvík	stud-horse and pig grazing in Hraunsdalur
DI 2, 738	1340	Þykkvabær monastery	sow with four piglets
DI 3, 198	1363	Hrafnseyri, Arnarfjörður	calf and pig grazing in Sviðningsdalur, Dynjandaland
DI 3, 324	1377	Holt, Önundarfjörður	three-winter old sow with three piglets
DI 3, 515	1394	Árskógur, Eyjafjörður	stud-horse and pig grazing in Fagrahlið, Eyjafjörður
DI 4, 145	1397	Hrafnseyri, Arnarfjörður	calf and pig grazing in Sviðningsdalur, Dynjandaland
DI 4, 144	1397	Mýrar, Dýrafjörður	common pasture for pigs and stud-horses in Mjóadalur
DI 6, 472	1483	Reykhólar, Westfjörðs	two old pigs and six piglets brought to Reykhólar
DI 7, 184	1493	Staðarfell, Fellsstrandarhreppur	three pigs and a sow with nine piglets
DI 10, 339	1537	Bishop Ögmundur of Skálholt	record of payment included five pigs
DI 12, 596	1553	Viðey, Kollafjörður	account record included four pigs
DI 15, 571	late 16th century	Hóll, Bolungarvík	stud-horse and pig grazing in Hraunsdalur
DI 15, 579	late 16th century	Hrafnseyri, Arnarfjörður	calf and pig grazing in Sviðningsdalur, Dynjandaland
Jarðabók 7, 15–16	Sept. 1710	Hrafnseyri, Arnarfjörður	calf and pig grazing in Sviðningsdalur, Dynjandaland, grassless and barren

DI = Jónsson et al. eds (1857–1950), *Jarðabók* = Magnússon and Vídalín (1980–1988)

If we look at the geographical locations, we discover that most of these documents refer to the Westfjörðs of Iceland, an area known for steep valleys and limited agricultural land and grazing. This region is also near fruitful fishing grounds. It might be the case that the wealth produced from fishing enabled the rearing of pigs, which can be fed on fish waste.

While there is physical evidence in the form of zooarchaeological finds for the existence of pigs into the fourteenth century, the documentary sources show pigs were still present on wealthier church-farms in the mid-sixteenth century. The other source of evidence available from this time is the observational writings about Iceland by Icelanders. Towards the end of the sixteenth century, Bishop Oddur Einarsson of Skálholt, the southern diocese, stated in his writings that pigs were still reared in Iceland, found in deep valleys and areas of marshland, kept inside and fed leftovers and hay during cold weather (Einarsson 1971, 101–102). He did not give the locations of the farms that kept pigs, but based on the church-charters, the farms probably included the Westfjörðs. However, just over a century later when *Jarðabók* was compiled, no pigs were recorded on any Icelandic farm. The writings of Eggert Ólafsson and Bjarni Pálsson in the mid-eighteenth century further emphasise the lack of native-reared pigs in Iceland by stating that foreign merchants imported pigs and pork for their own consumption (Ólafsson and Pálsson 1805, 29).

Pigs were never numerous in Iceland even during its settlement, but as the centuries passed, their numbers became fewer and they were confined to wealthier church-farms. There is the possibility that wealthier secular farms also reared pigs, but we lack evidence about these farms. Pig rearing did continue through to the late sixteenth century, but sometime between the last decades of the sixteenth century and the start of the eighteenth century, pigs ceased to be raised on Icelandic farms. Several explanations can be proposed, ranging from a lack of suitable environments to an unprofitable return on resources invested in their husbandry.

Goats appear to have fared better in Iceland, though sources charting their husbandry in Iceland are still as scarce as the sources for pigs. As with pigs, the zooarchaeological evidence describes how goats were more common during the settlement of Iceland in the ninth and tenth centuries, before becoming absent on 'normal' farms by the mid-tenth century, and then rare on all farms by the thirteenth century. At the famous site of Hofstaðir, Mývatn, in north-east Iceland, the ninth- and tenth-century zooarchaeological evidence shows that goats were exploited for milk but increased relative to sheep over time, from a ratio of goat to sheep of approximately 1:7 to 1:5. Meanwhile, at the neighbouring farm of Sveigakot, the goat to sheep ratio drastically decreased from about 1:3 to 1:13 over the same timeframe in the ninth and tenth centuries (McGovern 2009, 204–205). At the late medieval and early modern farm of Stóraborg, Eyjafjallasveit, the goat to sheep ratio was approximately 1:9, showing that goats were still being consumed, and therefore reared, but not in the same numbers as sheep. The identification of a single goat bone at Reykholt, Borgarfjörður and at Viðey, Kollafjörður demonstrates that goats were not commonly consumed and the single bones may represent cuts of meat or items brought to these farms (Sveinbjarnardóttir et al. 2007, 201; Amorosi 1996, 407). A single goat horn core in a collection of 13,000 animal remains recovered from the late-fourteenth to early-fifteenth century Skriða monastery, Fljótsdalur, eastern Iceland, has been explained as raw material brought to the site for tool manufacturing (Baldursdóttir et al. 2017, 2–3). A mid-eighteenth-century context at Skálholt only produced two goat bones and 106 sheep bones with 515 caprine bones mainly from meat-bearing parts of the carcass, indicating that although this site was wealthy and had access to resources goats did not feature regularly in the inhabitants' diet (Hambrecht 2011, 110). The recovery of individual or small numbers of goat horn cores, as at Skriða, may indicate raw material or an artefact brought to the site, whereas other goat bones, especially from meat-rich body parts such as at Skálholt, would indicate consumption of goats at a site.

Larger numbers of goat remains have been recovered from late medieval to early modern sites, such as at Svalbarð, Þistilfjörður, north-east Iceland, Stóraborg, southern Iceland and the bishopric of Skálholt, southern Iceland. From contexts dated from the fourteenth to eighteenth centuries at Svalbarð, five goat and 39 sheep bones were identified in addition to 127 bones classified as caprines. In similarly dated contexts at Stóraborg, there were 29 goat and 273 sheep bones identified, and 1139 categorised as caprines (Hambrecht 2009, 20). These numbers give some idea about the scale of difficulty involved when distinguishing goat from sheep. Nevertheless, these clearly identified goat bones can tentatively be seen as evidence for the consumption, and possibly rearing, of goats in Iceland between the fourteenth and eighteenth centuries. The southern location of Stóraborg, however, away from the documented goat herds of the north and the small number of goat bones do raise questions about whether goats or cuts of goat-meat were brought to the site and not reared on the farm.

The zooarchaeological evidence thus demonstrates that goats were more common in the earlier centuries of Iceland's history and were being consumed in relatively small numbers at some farms around Iceland. They continued to be consumed in ever smaller quantities, but goat rearing was still continued in Iceland.

Our other approach for examining goat rearing in Iceland is the written sources, and it is to them that we now turn for indications of how goats were farmed. The Sagas of Icelanders, set in the ninth to eleventh centuries, contain infrequent details that provide explanations for the rearing of goats in Iceland. In the widely known *Brennu-Njáls saga*, goat-skins are twice mentioned as items of clothing, one of which was used in ritual activity hinting at a supernatural aspect of goats (Sveinsson ed. 1954, XII chapter 133, 347 and chapter 12, 37). A saying in another saga set in the earlier centuries expresses that a person who seeks wool from goats is foolish, emphasising that wool was not generated by goats in Iceland (Vilmundarson and Vilhjálmsson eds 1991, XIII chapter 1, 331). Goats were also housed, and a passing reference in one of the Sagas of Icelanders during a search of a goat-shed for a wanted man shows that goats were conceivably sheltered in the same structure as other livestock (Jóhannesson ed. 1950, XI chapter 19, 276). A unique billy-goat was allowed to graze the valuable home-field of Bergþórshvoll in southern Iceland, showing it was conceivable for goats to be kept in this area though grazing the home-field was a privilege reserved for special animals (Sveinsson ed. 1954, XII chapter 41, 106, see also Sveinsson and Thorðarson eds 1935, IV chapter 63, 172). In a supernatural saga episode, two grey lambs and two grey kids are stolen by a troll and a shape-shifter, and this illustrates the similarities in practices for sheep and goats as they were kept together in a pen (Vilmundarson and Vilhjálmsson eds 1991, XIII chapter 12, 189). The inclusion of goats in these sagas depicting the ninth to eleventh centuries differs from the lack of goats in the Saga of the Sturlungs, which recount events of the twelfth to fourteenth centuries. The Árnessýslu price-list from c.1200 does not mention goats, indicating that by this date, goats had already begun to become geographically limited and were no longer found in this area of south-west Iceland (Jónsson et al. eds 1857–1950: 1, 316–317). Goat farming, however, did continue in Iceland as seen in the church-charters, Table 2.2.

Table 2.2 shows that there was physical evidence of goats and their historical grazing areas. As the church-charters are available for the whole of Iceland, we can be more certain of geographical preferences, and goat rearing seems to have been more concentrated in northern Iceland in Eyjafjörður and the neighbouring valleys. Indications of the importance of recording grazing rights is found at Laufás, Eyjafjörður, where the continuous recording of pasture for ewes and goats is found in charters centuries apart. The repeated recording of grazing rights at Laufás emphasises

Table 2.2 Summary of documentary evidence for the presence of goats

Source	*Date*	*Location*	*Description*
DI 1, 522	1238	Gufudalur, Gufufjörður, Westfjörðs	place-name of Geithúslækur recorded (geithús = goat-house)
DI 2, 439	1318	Háls, Fnjöskadalur	farm owned 80 ewes with goats
DI 2, 447	1318	Laufás, Eyjafjörður	farm had access to grazing for as many ewes and female goats as desired for six weeks
DI 2, 450	1318	Gnúpufell, Eyjafjörður	one ***kúgildi*** (value of an average milking cow) of goats
DI 3, 566	1394	Laufás, Eyjafjörður	farm had access to grazing for as many ewes and female goats as wished for six weeks
DI 5, 120	1454	Reykir, Ólafsfjörður	a goat-house is used as a land-marker in a letter of sale
DI 5, 127	1454	Hofteig, Jökuldalur	goat-house
DI 5, 266	1461	Laufás, Eyjafjörður	farm had access to grazing for as many ewes and female goats as desired for six weeks
DI 6, 552	1485	Vatnsleysa, Fnjóskadalur	land sale includes goats
DI 6, 581	1486	Munkaþverá monastery, Eyjafjörður	Abbot Einar gave his relative gifts including three goats
DI 9, 309	1525	Illugastaðir, Fnjóskadalur, owned by Munkaþverá	record of possessions lists eight female goats, two winter-old female goats and three male goats
DI 9, 329	1525	Laufás, Eyjafjörður	farm had access to grazing for as many ewes and female goats as desired for six weeks
DI 11, 855	1550	Hólar, northern bishopric	20 female and male goats, and two females with four kids
DI 13, 357	1558	Grazing stipulation	goats can graze in the woods but must not damage others' land
Jarðabók	1703	Þingeyjarsýslu, NE Iceland	818 goats recorded in all of Iceland, along with 278,994 sheep, 35,860 cattle, and 26,909 horses

DI = Jónsson et al. eds (1857–1950), *Jarðabók* = Magnússon and Vídalín (1980–1988)

the value of these rights and the perceived need to copy them from one charter to another with the same conditions to prove the historical rights of the farm.

The rearing of goats continued into the late sixteenth century and their numbers even increased, such as at Hólar, the northern bishopric, which in 1550 owned 20 female and male goats, and two females with kids, giving a total of 22 adult goats, whereas by 1586 there was 32 adult goats consisting of 20 female goats, 13 kids, 10 gelded male goats, and 2 male goats (Júliusson and Jónsson 2013, VI, 107; Jónsson et al. eds 1857–1950, 11, 855). The documents before the eighteenth century are almost entirely concerned with church-farms rather than secular farms. As such, larger, wealthier church-farms in areas known to have woodland appear to have had access to woodland grazing and thereby had the resources and wealth to keep sizable herds of goats. As the ten gelded goats at Hólar in 1586 could not have been used for breeding and did not produce milk, they must have been kept for skins and meat. Oddur Einarsson, the bishop of Skálholt mentioned above, in his 1589 book about Iceland praised goats for milking well and having good skins (Einarsson 1971, 101). He wrote, however, that goats were kept in the scarce wooded areas, which is supported by the 1558 grazing stipulation permitting goats to graze in woods on the condition that they do not damage others' land (Jónsson et al. eds 1857–1950: 13, 357). We, thus, have an image of goat rearing by the sixteenth century being confined to areas that had woodland which was itself a scarce resource. Half a century after Oddur Einarsson, his son Bishop Gísli Oddsson of Skálholt also compiled his own writings about Iceland, in which goats are named among Icelandic livestock, demonstrating that the species was known, and were still present on some Icelandic farms (Oddsson 1942, 96). The number of goats in Iceland at this time is uncertain and it is unclear if the total goat population in Iceland was increasing or decreasing. There is some speculation about whether the 818 goats recorded in *Jarðabók* were part of a downward trend in goat rearing as less and less goats were recorded in Iceland between *Jarðabók* and the twentieth century (Júliusson and Jónsson 2013, VI, 108; Thoroddsen 1908–1922, VI, 71–72). What is certain is that goats were reared throughout Iceland's history but formed only a small part of the livestock economy.

From the physical and written evidence it is clear that pigs and goats were brought to Iceland and bred, though in relatively fewer numbers than cattle and sheep. By the thirteenth century it is clear that pigs and goats had lost out to cattle and sheep as the significant livestock species. However, that is where the similarities between pigs and goats end. Pigs decreased in number and geographical distribution, seemingly reared for the longest in the Westfjörðs, before finally fading from Icelandic farms in the late sixteenth century, only to remain fossilised in sagas and charters. Merchants that came to Iceland brought pork and live pigs with them to rear and slaughter in trading places, not for rearing on farms. Goats, on the other hand, continued to be reared with viable breeding herds in several valleys in the north and north-east of the country. Though we only have sparse records from church-farms throughout the medieval period, *Jarðabók* at the beginning of the eighteenth century demonstrates that goats were still present on Icelandic farms, but pigs were already absent.

Discussion

When considering the differing fates of pigs and goats in Iceland, two aspects of their husbandry must be examined: fodder and products. Pigs, being omnivores, are useful in fledgling settlements as they can eat almost anything, including leftovers and waste from fish processing. Pigs also favour woodland, which Iceland had increasingly little of, and their foraging behaviour damaged the fragile Icelandic soils. As early as the twelfth century, however, *Grágás* was stipulating that pigs without nose-rings were restricted to their owner's land and could be killed if found trespassing (Dennis et al. 2000, K.207, 139). *Grágás* is thus indicating an early awareness among the Icelanders of pigs' detrimental effect on vegetation and soil. Bishop Oddur Einarsson in the late-sixteenth century stated that pigs needed to be housed during winter (Einarsson 1971, 101–102). Pigs, therefore, were unsuitable for the Icelandic environment, and Iceland was unsuitable for pigs because of the harsh winter and the lack of the diverse woodlands found in mainland Europe. Furthermore, while pigs are able to reproduce quickly, were a ready source of meat, and helped open up wooded areas during the settlement, they had a limited return for the time and resources invested in them. In Iceland, other livestock could produce the same products with greater ease and less risk to the environment. For example, sheep for wool, meat, and milk grazed uplands during the summer, and only required a labour investment when the sheep were driven down to the farms in the autumn, and then for shearing and lambing. Allowing pigs to graze freely was also not a legally available option.

Goats, in contrast, produce a similar range of products as sheep, except for wool, and can produce milk for longer than sheep, even into the winter. Indeed, where we have information about goats they appear to be mainly nanny-goats, pointing to the use of goats for milk production. To understand the place of goats on Icelandic farms, we need to step back and view the wider context. The church-charter evidence shows a greater variability in farming strategies from the fourteenth century onward, but with an overall shift towards wool-bearing sheep. This shift is also noted in the archaeological record from the thirteenth century. It appears, therefore, that on farms generally there was a move towards wool production for woollen cloth. *Vaðmal*, a woollen cloth, was a mainstay of the Icelandic domestic economy throughout the medieval period, and a significant export good to Norway and the British Isles into the late medieval period (Hayeur Smith 2015). While goats produce a range of products, they lost out because they did not generate wool, which was so important to the economy, international trade, and ultimately, Iceland remaining in contact with Europe. While goats could graze grasses, the practice put them in direct competition with the wool-producing sheep. Goats prefer woodlands and shrubs, and in *Jarðabók*, goats are recorded on farms with rights to such land, but they became geographically restricted as woods and shrubland became depleted over the years. Goats were thus restricted to grazing in scarce wooded areas so as not to put them into competition with wool-producing sheep. At the same time, the number of goats would have needed to be controlled to protect the valuable woodland resources, hence a balance had to be struck between the number of goats and the availability and preservation of woodland. Deforestation is thought to have begun in the lowlands from the time of settlement in the ninth century, while woodlands were preserved in some upland areas until the eighteenth and nineteenth centuries when the rate of deforestation increased (Vésteinsson and Simpson 2004, 185). Goat owners would have required rights, or at least access, to these woods as well, which would be a deciding factor on whether goats were owned in the first place. In *Jarðabók*, for example, many of the secular farms in Fnjóskadalur that had goats owned or had rights to such land (Júliusson and Jónsson 2013, 107).

The rarity of pigs and goats would have made their ownership something of a symbol of wealth or high status. The keeping of relatively unproductive livestock was a luxury that few could afford, especially as the climate became more unpredictable and human population losses shook the labour force. Pigs were not diverse enough in their products to warrant the effort and resources necessary for their keeping, and their foraging was damaging to the Icelandic environment. Goats were out-competed by wool-producing sheep, but continued to be reared in Iceland because they had a variety of products and some grazing was still available for them.

As the settlers of the North Atlantic moved westward, they brought with them the same livestock species, including pigs and goats. What these settlers and their descendants did after they established themselves differed for each region. A brief comparison with Iceland's two closest neighbours illustrates how these societies each adapted to their own conditions and how these two species succeeded and lost in the North Atlantic.

In Greenland goats were relatively more common than elsewhere in the North Atlantic. They were just as numerous as sheep on larger farms, and on smaller farms accounted for most of the caprine species throughout the two Norse settlements (Mainland and Halstead 2005, 117; Smiarowski et al. 2017, 152). Both goats and sheep in Greenland were reared for milk and meat, with little emphasis on wool-production. Goats also appear to have been better suited to the Greenland environment (Mainland and Halstead 2005, 116). Unlike in Iceland and the Faroes, Greenland may have been unable to generate surplus wool for the export market because of the need to restock their flocks after harsh winters, and thus, wool generation was not constant year-to-year (Dugmore et al. 2013, 446). Greenlanders instead depended heavily on walrus ivory for international trade, and the reliance on one product made them vulnerable to shifting trends in medieval Europe and changes in the walrus population that could have eventually contributed to their own extinction (Star et al. 2018, 6; Smiarowski et al. 2017, 155).

Pigs appear to have been initially common on larger farms in Greenland but, as in Iceland, fell out of favour and became scarce by the eleventh century and rare by the twelfth century according to the archaeological record (Brewington et al. 2015, 1677; McGovern et al. 2007, 30). Unfortunately, there is not the wealth of documentary sources available for Greenland as there is for Iceland and the Faroe Islands and so we cannot rely on church-charters or observational writings for further information. Nevertheless, a certain end-date for all livestock farming is the mid-fifteenth century when both Norse settlements had themselves become extinct (McGovern 1991, 94).

The Faroe Islands are known for their sheep, but domestic animals make up a small proportion of the overall zooarchaeological collections, being outnumbered by the remains of birds and fish. Among the livestock species, goats are barely detectable at Undir Junkarinsfløtti, dating from the ninth to thirteenth century. For example, there was only one bone positively identified as goat from a context dated to the eleventh and twelfth centuries (Church et al. 2005, 189). A single goat bone does not indicate goat farming, so caution must be exercised when drawing conclusions. The treeless environment of the Faroe Islands would have deprived goats of woodlands to utilise, but it is also the case that further evidence of goats may be hidden in the combined 'caprine' category (Baldursdóttir et al. 2017, 3). The preference for sheep can be linked to the importance wool played in the economy of the Faroe Islands, and so goats lost out to sheep in both fodder and products.

On the other hand, there is more evidence for the rearing of pigs in certain areas of the Faroe Islands, in the form of place-names that include elements referring to pigs. Research on these names suggests that pigs were closely supervised and moved along defined routes on the northern islands before the Reformation, and isotopic analysis reveals that pigs had similar feeding practices as cattle (Arge et al. 2009, 19; 29). Domestic species, however, constitute a small part of the zooarchaeological collections, approximately 5% of the total Undir Junkarinsfløtti collection, and pigs represent only a fraction of the domestic animal bones (Brewington et al. 2015, 1678–1679). Pigs were therefore only ever a small part of Faroese farming. The protection of small-scale barley cultivation and puffin nesting sites from pigs' rooting habits has been used to explain the reduction of pig rearing on the Faroes (Brewington et al. 2015, 1679). The documentary sources testify that in the sixteenth to eighteenth centuries pig keeping was an uncommon practice with pigs confined to some farms (Arge et al. 2009, 19–20). It appears that the Faroese were willing to give up, or significantly restrict, pigs in order to protect and preserve their arable farming, and more importantly the puffins on which they relied for a larger proportion of their diet. As in Iceland, pigs required more resources and labour than their products warranted, as well as posing a threat to the environment. However, later written sources do indicate that pigs were present in the Faroe Islands into the early modern period, unlike in Iceland.

Conclusions

Adaptation to the Icelandic environment and climate have continued throughout the human occupation of Iceland, and changes in livestock numbers and proportions can be seen as reflecting wider economic conditions. To understand any adaptation we need to consider the winners and losers over the long-term, otherwise we risk generating short-sighted explanations. As seen in Iceland, changes in livestock husbandry was a balancing game: fodder needs had to be matched with the pay-off of products. After the initial settlement and especially with the onset of climatic change that stressed the agricultural system, livestock that damaged the environment or were too expensive in terms of labour and resources could no longer be afforded. From the zooarchaeological evidence it appears that smaller farms were the first to cease pig farming. The written sources show that by the mid-sixteenth century, even the wealthier church-farms did not rear them. Goats, in contrast, appear to have fared the same as pigs based on the archaeological collection, but the written sources demonstrate they were kept in small numbers, and restricted to certain areas with woodlands so as not to compete with the valuable wool-producing sheep instrumental in maintaining international trade connections. *Jarðabók* testifies that goats, while scarce, did continue to be farmed on smaller farms and were not confined to larger, wealthier farms, as had been the case for pigs. Only by taking a long-term approach to these species are we able to differentiate their fates.

The same livestock species were introduced across the North Atlantic. The initial settlers and the following generations adjusted the relative proportions of species they farmed, in some cases ceasing to farm certain livestock, so as to make the most of their labour input and preserve their environmental resources. *Grágás* demonstrates that this was the case in the early centuries of Icelandic history, as does the 1298 *Sauðabrævið*, 'sheep-letter,' that set out resource management on the Faroe Islands. These people were aware and willing to react to changes around them in order to survive and maintain links with the wider world.

From the comparison of Iceland and its two closest geographical neighbours, we can see the extent of adaptation. Pigs were clearly ill-suited to the North Atlantic *milieu*, and the cessation of pig farming occurred both for environmental and socio-economic reasons. Pigs were unproductive compared with other livestock, and an expense that could not be afforded even by the elites. Pigs were useful during the establishment of these societies but required too much labour to supervise them, as well as too many resources to feed and house them. The evidence from the Faroe Islands illustrates how closely controlled pigs needed to be and that they had the potential to damage the more valuable arable farming and bird nesting sites. Goats, on the other hand, followed a different path. In Iceland and the Faroes, goats became rare, though the timing and geographical distribution of their scarcity differed. The written sources from Iceland demonstrate an understanding of the advantages of goat rearing, but on balance sheep helped to maintain Iceland's place in international trade. Greenland stands in stark contrast to the other regions, as goats remained prominent in the archaeological record. The sole focus of Greenlanders on valuable walrus ivory restricted their attractiveness to overseas merchants, whereas Iceland and the Faroes shifted to producing wool products that were desired in the markets of medieval Europe.

This chapter has shown that we need to consider the species that lost out, because they give us another means to examine the ways farming responded to changing conditions. Decisions about livestock husbandry were not uniform across these islands or within them. Rearing livestock depended on many factors: the availability of fodder and labour, the return on the resources invested in keeping these animals, the suitability of the climate to the livestock, and the environmental sustainability of husbandry practices. All three regions considered in this discussion—Iceland, Greenland, and the Faroe Islands—demonstrate adaptation to these factors to varying degrees, as shown by the different outcomes for pig and goat husbandry.

Acknowledgments I would like to thank the participants of the Medieval Animal Data Network (MAD) and the Domestic Animals in the Viking Age (DAVA) workshops, both held in November 2017, for their feedback on earlier versions of this chapter.

References

Primary Sources

Benediktsson, Jakob ed. 1986. *Íslendingabók-Landnámabók*. Íslenzk fornrit vol. I. Reykjavík: Hið íslenzka fornritafélag.

Dennis, Andrew, Peter Foote, and Richard Perkins transl. 2000. *Laws of Early Iceland: Grágás II*. Winnipeg: University of Manitoba Press.

Einarsson, Oddur. 1971. *Íslandslýsing. Qualiscunque description Islandiae (1589)*, translated by Sveinn Pálsson. Reykjavík: Bókaútgáfa Menningarsjóðs.

Jóhannesson, Jón ed. 1950. *Austfirðinga sögur*. Íslenzk fornrit vol. XI. Reykjavík: Hið íslenzka fornritafélag.

Jóhannesson, Jón, Magnús Finnbogason, and Kristján Eldjárn eds. 1946. *Sturlunga saga* I. Reykjavík: Sturlunguútgáfan.

Jónsson, Þorlákur, Ólafur Daviðsson, Ólafur and Jón Þorkelsson eds. 1857–1950. *Diplomatarium Islandicum: Íslenzkt fornbréfasafn, sem hefir inni að halda bréf og gjörninga, dóma og máldaga, og aðrar skrár, er snerta Ísland eða íslenzka menn* I–XVI. Reykjavík: Félagsprentsmiðju.

Magnússon, Árni, and Páll Vídalín. 1980–1988. *Jarðabók Árna Magnússonar og Páls Vídalíns*. Vestmannaeyjasýsla, Rangárvallasýsla, vols. I–XI, edited by Melsteð, Bogi Th., Finnur Jónsson, Björn K. Þórólfsson, Jakob Benediktsson, and Gunnar F. Guðmundsson. Reykjavík: Sögufélagið.

Oddsson, Gísli. 1942. *Íslenzk annálabrot [Annalium in Islandia farrago] og Undur Íslands [De mirabilibus Islandiæ]* 1638, translated by Jónas Rafnar. Akureyri: Þorsteinn M. Jónsson.

Ólafsson, Eggert, and Bjarni Pálsson. 1805. *Travels in Iceland by Eggert Ólafsson and Bjarni Pálsson (1752–1757)*, translated. London: R. Phillips.

Schulman, Jana K. 2010. *Jónsbók: The Laws of Later Iceland*. Saarbrücken: AQ-Verlag.

Sveinsson, Einar Ólafur. 1939. *Vatnsdæla saga*. Íslenzk fornrit vol. VIII. Reykjavík: Hið íslenzka fornritafélag.

Sveinsson, Einar Ólafur.1954. *Brennu-Njáls saga*. Íslenzk fornrit vol. XII. Reykjavík: Hið íslenzka fornritafélag.

Sveinsson, Einar Ólafur, and Matthias Thorðarson. 1935. *Eyrbyggja saga*. Íslenzk fornrit vol. IV. Reykjavík: Hið íslenzka fornritafélag.

Vilmundarson, Þórhallur, and Bjarni Vilhjálmsson. 1991. *Harðar saga*. Íslenzk fornrit vol. XIII. Reykjavík: Hið íslenzka fornritafélag.

Secondary Sources

Amorosi, Thomas. "*Icelandic Zooarchaeology: new data applied to issues of historical ecology, palaeoeconomy and global change.*" PhD diss., City University of New York, 1996.

Arge, Símun V., Mike Church, and Seth Brewington. "Pigs in the Faroe Islands: an Ancient Facet of the Islands' Paleoeconomy". *Journal of the North Atlantic* 2 (2009). 19–32.

Baldursdóttir, Birna Kristín, Albína Hulda Pálsdóttir, and Jón Hallstein Hallsson. "Geitfé á Íslandi – uppruni, staða og framtíðararhorfur". *Skrína* 3, vol. 2 (2017). 1–15.

Brewington, Seth, Megan Hicks, Ágústa Edwald, Árni Einarsson, Kesara Anamthawat-Jónsson, Gordon Cook, Philippa Ascough, Kerry L. Sayle, Símun V. Arge, Mike Church, Julie Bond, Steve Dockrill, Adolf Fríðriksson, George Hambrecht, Árni Daniel Júlíusson, Vidar Hreinsson, Steven Hartman, Konrad Smiarowski, Ramona Harrison, and Thomas H. McGovern. "Islands of change vs. islands of disaster: Managing pigs and birds in the Anthropocene of the North Atlantic." *The Holocene* 25, no. 10 (2015). 1676–1684.

Church, Mike, Símun V. Arge, Seth Brewington, Thomas H. McGovern, Jim Woollett, Sophia Perdikaris, Ian T. Lawson, Gordon Cook, Colin Amundsen,

Ramona Harrison, Yekaterina Krivogorskaya, and Elaine Dunbar. "Puffins, Pigs, Cod and Barley: Palaeoeconomy at Undir Junkarinsfløtti, Sandoy, Faroe Islands." *Environmental Archaeology* 10 (2005). 179–197.

Dugmore, Andrew J., Thomas H. McGovern, Richard Streeter, Christian Madsen, Konrad Smiarowski, and Christian Keller. "'Clumsy solutions' and 'elegant failures': Lessons on climate change adaptation from the settlement of the North Atlantic islands." In *A Changing Environment for Human Security: Transformative Approaches to Research, Policy and Action*, edited by Linda Sygna, Karen O'Brien, and Johanna Wolf, 435–451. London: Routledge, 2013.

Edvardsson, Ragnar. "Statistical Analysis of the 1703–1712 Land Register: Four Districts in the North West of Iceland." In *Current Issues in Nordic Archaeology: Proceedings of the 21st Conference of Nordic Archaeologists 6–9 September 2001 Akureyri Iceland*, edited by Gardar Guðmundsson, 189–192. Reykavík: Félag íslenskra fornleifafræðinga, 2004.

Hambrecht, George. (2011). "*Faunal Analysis of the Early Modern Bishop's Farm at Skalholt, Arnessysla Iceland.*" PhD diss., City University of New York, 2011.

Hambrecht, George. "Zooarchaeology and the Archaeology of Early Modern Iceland." *Journal of the North Atlantic* 1 (2009). 3–22.

Harrison, Ramona. (2013). *"World Systems and Human Ecodynamics in Medieval Eyjafjörður, North Iceland: Gásir and its hinterlands."* PhD diss., City University of New York, 2013.

Harrison, Ramona. "Connecting the Land to the Sea at Gásir: International Exchange and Long-Term Eyjafjörður Ecodynamics in Medieval Iceland." In *Human Ecodynamics in the North Atlantic: A collaborative model of humans and nature through space and time*, edited by Ramona Harrison, and Ruth Maher, 117–136. London: Lexington Books, 2014.

Harrison, Ramona, Howell M. Roberts, and W. Paul Adderley. "Gásir in Eyjafjörður: International exchange and local economy in medieval Iceland." *Journal of the North Atlantic* 1 (2008). 99–119.

Hayeur Smith, Michèle. "Weaving Wealth: Cloth and Trade in Viking Age and Medieval Iceland." In *Textiles and the Medieval Economy: Production, Trade, and Consumption of Textiles, 8th–16th Centuries*, edited by Angela Ling Huang, and Carsten Jahnke, 23–40. Oxford: Oxbow, 2015.

Júlíusson, Árni Daníel, and Jónas Jónsson. 2013. *Landbúnaðarsaga Íslands* I–IV. Reykjavík: Skrudda.

Karlsson, Gunnar. (2003). *The History of Iceland*. London: Hurst Co. Ltd.

Mainland, Ingrid, and Paul Halstead. "The Economies of Sheep and Goat Husbandry in Norse Greenland." *Arctic Archaeology* 42, no. 1 (2005) 103–120.

McGovern, Thomas H. "The Archaeofauna." In *Hofstaðir: Excavations of a Viking Age Feasting Hall in North-eastern Iceland*, edited by Gavin Lucas, 168–252. Reykjavík: Institute of Archaeology, 2009.

McGovern, Thomas H., Orri Vésteinsson, Adolf Friðriksson, Mike J. Church, Ian T. Lawson, Ian A. Simpson, Árni Einarsson, Andy Dugmore, Gordon Cook, Sophia Perdikaris, Kevin Edwards, Amanda M. Thomson, W. Paul Adderley, Anthony Newton, Gavin Lucas, Ragnar Edvardsson, Oscar Aldred, and Elaine Dunbar. "Landscapes of Settlement in Northern Iceland: Historical Ecology of Human Impact and Climate Fluctuation on the Millennial Scale." *American Anthropologist* 109, no. 1 (2007). 27–51.

McGovern, Thomas H., Sophia Perdikaris, Árni Einarsson, and Jane Sidell. "Coastal connections, local fishing, and sustainable egg harvesting: patterns of Viking Age inland wild resource use in Mývatn district, Northern Iceland." *Environmental Archaeology* 11, no. 2 (2006). 187–205.

McGovern, Thomas H., Sophia Perdikaris, and Clayton Tinsley. "Economy of Landnám: The Evidence of Zooarchaeology." In *Approaches to Vinland: a conference on the written and archaeological sources for the Norse settlements in the North-Atlantic region and exploration of America, the Nordic House, Reykjavík, 9–11 August 1999*, edited by Andrew Wawn, and Thorunn Sigurðardottir, 154–166. Reykjavik: Sigurdar Nordal Institute, 2001.

McGovern, Thomas H. "Climate, Correlation, and Causation in Norse Greenland." *Arctic Anthropology* 28, no. 2 (1991). 77–100.

Reitz, Elizabeth J., and Elizabeth S. Wing. 2008. *Zooarchaeology*. Cambridge: Cambridge University Press.

Sigurðsson, Jón Viðar. 1999. *Chieftains and Power in the Icelandic Commonwealth*, translated by Jean Lundskær-Nielsen. Odense: Odense University Press.

Simpson, Ian A., Andrew J. Dugmore, Amanda Thomson, and Orri Vésteinsson. Crossing the thresholds: human ecology and historical patterns of landscape degradation. *Catena* 42 (2001). 175–192.

Smiarowski, Konrad, Ramona Harrison, Seth D. Brewington, Megan Hicks, Frank Feeley, Céline Dupont-Hébert, Brenda Prehal, George Hambrecht, James Woollet, and Thomas H. McGovern. "Zooarchaeology of the Scandinavian settlements in Iceland and Greenland: diverging pathways." In *The Oxford handbook of Zooarchaeology*, edited by Umberto Albarella, Mauro Rizzetto, Hannah Russ, Kim Vickers and Sarah Viner-Daniels, 147–163. Oxford: Oxford University Press, 2017.

Star, Bastiaan, James H. Barrett, Agata T. Gondek, and Sanne Boessenkool. "Ancient DNA reveals the chronology of walrus ivory trade from Norse Greenland." *Proceedings of the Royal Society B* 285 no. 1884 (2018). 1–9.

Sveinbjarnardóttir, Guðrún, Egill Erlendsson, Kim Vickers, Thomas H. McGovern, Karen B. Milek, Kevin J. Edwards, Ian A. Simpson, and Gordon Cook. "The palaeoecology of a high status Icelandic farm." *Environmental Archaeology* 12, no. 2 (2007). 187–206.

Thoroddsen, Þorvaldur. 1908–1922. *Lýsing Íslands* IV. Copenhagen: Íslenzka Bókmentafélag.

Vésteinsson, Orri, and Ian A. Simpson. "Fuel utilisation in pre-industrial Iceland. A micro-morphological and historical analysis." In *Current Issues in Nordic Archaeology: Proceedings of the 21st Conference of Nordic Archaeologists 6–9 September 2001 Akureyri Iceland*, edited by G. Guðmundsson, 181–188. Reykjavík: Society of Icelandic Archaeologists, 2004.

CHAPTER 3

Imperial Horse Policy and the Publication of Equine Veterinary Medicine Books in Ming China: A Case Study on *Yuanheng Liaomaji*

Zhexin Xu

Introduction

Horse veterinary care has had a long history in China (Meserve 1998; Cooke 2000; Buell et al. 2018). Horses are difficult to centrally manage in large breeding herds which makes veterinary care extremely important. Knowledge of healing horses was promoted by specialists during the Ming Dynasty who were trying to maintain the health of horse stocks for military and imperial purposes.

Among ordinary domestic animals, horse was ultimately ascribed the highest cultural status in many cultures, starting as a simple source of meat during prehistoric times until the Bronze age when it became a high-status token of human mobility, to a critical tool for maintaining military might

Z. Xu (✉)
Fachbereich Geschichte, Universität Salzburg, Salzburg, Austria
e-mail: zuzhexin@outlook.com

L. Bartosiewicz, A. M. Choyke (eds.), *Medieval Animals on the Move*, https://doi.org/10.1007/978-3-030-63888-7_3

and governance though inter-regional communication within the kingdom. This transition is mirrored in the way the early Ming Dynasty (AD 1368 to 1644) in China had already formulated an elaborate system of horse administration, an administrative system that was to last for centuries. Implementation of these horse policies influenced everyday life for a considerable proportion of the Chinese population while the increasing demand for horses also substantially shaped local societies in environmental, cultural, and economic terms. In this chapter, the growth of these complex administrative systems of horse management and breeding, influencing the emergence of veterinary scholarship, will be examined. These transformations in transit and communication are embodied in the popularity of Ming Dynasty veterinary literature through the example of *Yuanhen Liaomaji*, an equine veterinary encyclopedia published in the early seventeenth century.

Historical Background

The Ming Dynasty ruled China between AD 1368–1644. The dynasty emerged following the collapse of the Yuan Dynasty of Mongol extraction. It was to be the last imperial dynasty of ethnic Han Chinese. Overthrowing the governance of "alien nomads", a term frequently applied by the Han Chinese rebels to refer to their Mongolian masters, Ming rulers considered themselves the orthodox and legitimate successors of previous Chinese dynasties, holding the "heavenly mandate" to save China against "foreign rule". However, the Ming Dynasty still retained a lot of administrative and political institutions established by its former adversary, the Yuan Dynasty. Ming society was based on a rigid network structure of self-sufficient rural communities, sustaining a permanent ruling class represented by the army. Already during the early Ming Dynasty, almost ten percent of cultivated land was assigned to the military to sustain imperial soldiers and their families. In the meantime, the state also set aside horse pastures that were sometimes as much as five times larger than military farmlands (Robinson 2017, 298).

By the mid-fifteenth century, the Ming Dynasty ruled an area of well over six million square kilometers inhabited by a hundred million people. The Yangtze River Delta region, or Nanzhili ("southern directly ruled"), a province-level administrative district in which the Ming Dynasty's southern capital city of Nanjing was located, played a pivotal role in both the imperial horse breeding system and, closely connected to it, the imperial postal network, so vital to official administrative communications within

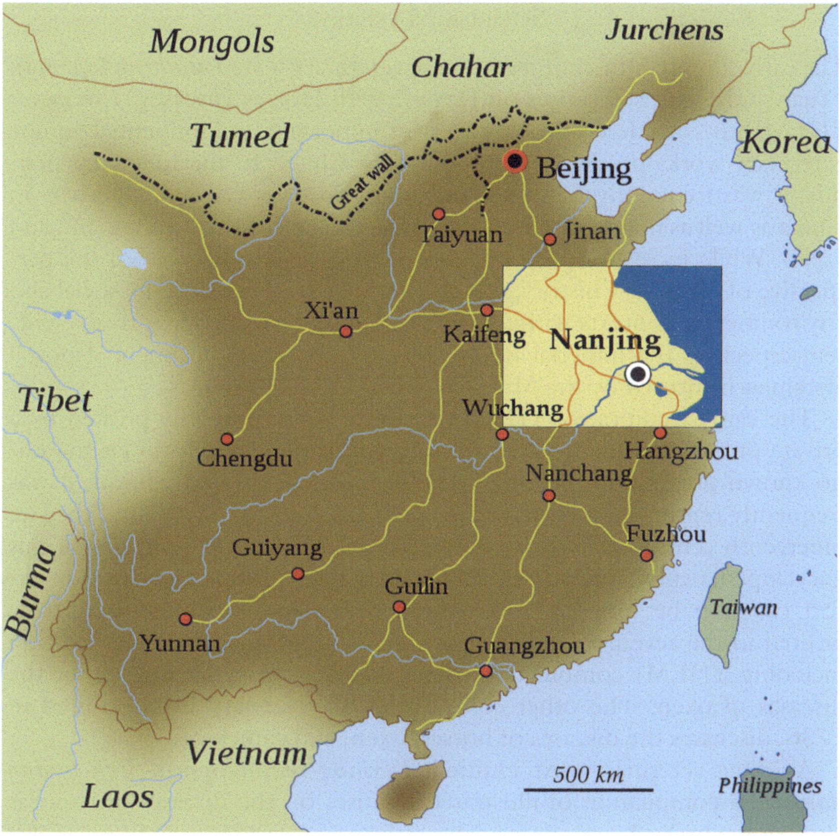

Fig. 3.1 The study area (highlighted) within Ming Dynasty China. Yellow lines between imperial and provincial capital cities generally illustrate the imperial postal system based on identification of postal stations from around 1580, most of which kept horses. Base map: Wikimedia Commons, image in the public domain after: https://commons.wikimedia.org/wiki/File:%C5%98%C3%AD%C5%A1e_Ming.png

the realm (Fig. 3.1). According to the chapter on the 1578 census in *Ming huidian*, the total population of the Ming Dynasty was 60,692,856, while Nanzhili alone possessed 10,502,651 inhabitants (Sheng Shixing et al., eds 1989, 123).

Source Material

The early seventeenth-century Chinese veterinary text, *Yuanheng Liaomaji* "Yuan and Heng's Treatises on Dealing with Horses Diseases" (hereafter YHLMJ), is considered one of the most important and influential equine veterinary works produced in pre-modern China. In the following periods, it was frequently reprinted and transcribed by various publishers in China as well as in other neighboring regions such as Korea, Vietnam, and Japan. While most previous studies of this text have been from the perspective of veterinary history, this chapter mainly focuses on the social and environmental contexts within which this equine veterinary text was written, especially the imperial horse policies as well as the imperial postal system as instituted by the Ming Dynasty.

The earliest edition of YHLMJ dates to 1608, primarily printed by a private publishing house, *Ruxian tang*, in Nanjing. While so far there is no known preserved copy of this 1608 woodcut-printed edition, it was frequently reprinted, even after the fall of the Ming Dynasty until the late nineteenth century. There are over 20 types of reprinted editions of this encyclopedia still preserved today. Besides the 1608 primary edition, its text can also be found in two edited compilations of veterinary texts printed in the seventeenth century. One of these is an illustrated edition including YHLMJ combined with another veterinary treatise about the diseases of oxen. The other compilation of veterinary texts, printed in 1736, discusses the diseases of horses, oxen, and camels.

Another veterinary text entitled *Tuxiang shuihuangniu jing hebing daquan*, a compilation of illustrated treatises on the diseases of oxen, is also known by Yu Yuan's name. The earliest preserved print edition of this latter book was printed in 1573 (Yu Yuan 1573), by a private publishing house, *Shulin jubao tang* in Nanjing (AA.47, Municipal Library of Yonezawa, Japan). Therefore, the putative authors, the Yu brothers, may have at least started writing their veterinary texts by the year 1573. It is also highly likely that an early manuscript or printing of YHLMJ already existed by this time, pre-dating the recognized 1608 edition by decades.

In this chapter, however, I will still treat the YHLMJ as an early seventeenth-century text, chiefly using a 1736 edition (437.4/432[1], Library of Nankai University) as well as a modern annotated version (Yu Benyuan and Yu Benheng [1736] ed. 1957) as the main reference.

The questions posed in this chapter are as follows:

- Why did YHLMJ appear in the late sixteen-century at the beginning of Ming rule in China?
- Why did various publishers keep on reprinting it?
- Who were the target audiences for this text?
- To what extent does the prevalence of this text relate to imperial horse policies and organization of the postal system at the time?

These research questions will be answered through contextualizing this treatise within the contemporary socio-political environment. The prevalence and spread of YHLMJ also represents a good example of inter-regional cultural and knowledge transfer in pre-modern China.

Authorship

As the first part of its title indicates, the authors of YHLMJ is commonly considered to be the Yu Yuan and Yu Heng brothers of the Yu Clan. Some scholars still argue that publishers at the time may have fabricated the title of this veterinary text and attributed it to the non-existent Yu Brothers in order to attract readers (Ramey and Rollin 2003, 29).

While there are still controversies about the authorship of YHLMJ, a 1608 preface to this book survived, written by a local intellectual-official named Ding Bin (1542–1633). It introduces the print of YHLMJ to the *Ruxian tang* publishing house and indicates that it is he who gave the titles for both veterinary texts on horse and cattle diseases using the names of the Yu brothers. According to Ding Bin, they were the authors of YHLMJ. He also briefly introduces the identities of the Yu brothers, as "reclusive veterinary surgeons that kindly composed these volumes and shared their knowledge to help the people and the country" (Yu Benyuan and Yu Benheng [1736] ed. 1957, 2).

When Ding Bin composed this preface, he had already earned the title of minister of the Ministry of Works (*Gongbu*) at Ming's Nanjing court, which was a relatively prestigious high-ranking governmental position (Yang Jialuo ed. 1975, 5829). Concerning his identity and status, the fact that he was willing to compose this preface implies that YHLMJ might have gleaned considerable attention in late Ming period China due to its particular value. However, it is uncertain whether Ding Bin knew the Yu brothers personally, as he does not further mention his connection to them in the preface.

Biographical information concerning the Yu Brothers is scarce in local gazetteers and dynastic chronicles of the time. One can only find scattered records in YHLMJ implying the identity of the authors. According to Ding Bin's preface and descriptions related to the management of horse breeding in imperial pastures, it is very likely that the Yu brothers were professional veterinary surgeons appointed by the imperial authority to assist local officers in managing horse breeding programs in imperial pastures. According to the date of the earliest print edition of YHLMJ, they were probably active in the Luzhoufu's Lu'an region (today Anhui Province, East China), around the second half of the sixteenth century.

Similarly to other veterinary as well as medical manuals produced in ancient China, YHLMJ is a so-called accumulated text. This style was rather common in traditional Chinese writings on medical and scientific knowledge. The author will frequently cite a variety of previous texts and descriptions concerning the same medical issues. Such books, thus, became a compilation of different texts to some extent, although in many cases, this style simply demonstrates writing skill aimed to make the text look more convincing. This feature of traditional Chinese writing is also visible in YHLMJ, as many chapters are intentionally written under the names of demi-god figures, such as Huangdi, Qibo, and Shihuang, legendary masters of medicine in the so-called primitive period (around 3000 BCE to 2000 BCE), according to traditional Chinese folklore. However, in YHLMJ it was actually the Yu brothers who composed most such entries in the name of those mystical figures. In addition, YHLMJ also cited and transcribed remedies and articles that had existed in previous veterinary texts, such as *Bole ji*, the "Essay collection of Bole", *Shihuang miji*, the "Secret collection of Shihuang", and *Mingyan fang* "Clearly effective remedies".

Structure, Content, and Audiences

The content of YHLMJ is divided into into four sections, named after the four seasons (Table 3.1). The first volume, *Chun juan* "Volume of Spring", discusses pragmatic methods for diagnosing horse diseases, as well as acupuncture therapy and skills connected to horse-judging. The second volume, *Xia juan* "Volume of Summer", listed 72 types of common horse diseases, accompanied by effective remedies used in curing them. The following *Qiu juan* "Volume of Autumn" and *Dong juan* "Volume of

Table 3.1 Structure of *Yuanheng Liaomaji*

Part	*Titles*	*Main content*
I	*12 Essays on Diagnoses*	Pulse examination; acupuncture; skin; standing position; sound bones; eyes and teeth.
II	*72 Types of Common Horse Diseases*	Effective remedies for curing 72 types of common horse diseases.
III	*Theoretical Analysis of 12 Types of Horse Diseases; Essays on 47 Types of Horses*	Theoretical discussion concerning the fundamental causes of different types of horse diseases.
IV	*5 must-know notes on Horse Breeding, Medicine Taking, and Acupuncture Therapy; Effective Remedies for 14 Categories of Horse Diseases*	Useful and effective instruction and guide for daily horse keeping.

Winter" respectively, present the causes of various horse diseases and offer pragmatic instruction connected to horse breeding.

YHLMJ contains descriptions of and remedies for over 300 types of horse diseases. These could be easily located, thanks to the book's index. Many of these entries were composed in the form of rhyming verse, enabling readers to easily memorize relevant texts. The 1608 edition of YHLMJ is also illustrated with 112 images. Thus, even a poorly educated peasant, was probably able to use this book to some degree. In comparison with other traditional Chinese veterinary texts, YHLMJ seems to be a pragmatic manual on horse husbandry, rather than a theoretical text aimed at professional readers.

It is worth noting that various private publishers reprinted YHLMJ frequently in the seventeenth and eighteenth century. There are over 30 types of preserved reprinted editions of YHLMJ produced during this period. These editions are currently held in archives in China and Japan (Feng Hongqian 2008, 29–30). Most were primarily printed by publishers located around Nanjing, such as *Ruxian tang*, *Songsheng tang*, and *Wangjuan lou*.

During this time, Nanjing and its neighboring areas were not only centres for printing and trading books: as will be illustrated in the following section, they also formed a crucial region in the Ming period imperial horse breeding system. The imperial pasture where the Yu brothers worked fell under the control of Nanjing as well. It is unsurprising that publishers concentrated in this region regularly produced pre-printed editions of YHLMJ, as local audiences must have had specific interests and needs connected to horses, creating a demand for veterinary manuals. In addition, the introduction and the application of an improved printing technique in Ming China enabled publishers and book merchants to produce and distribute veterinary texts on a large scale: printing became a mostly commercial activity aimed at earning profits. In this situation, a veterinary text was not simply a medium of knowledge transfer, but also a marketable commodity. The fact that YHLMJ has various reprinted editions implies that there was a constant and considerably strong market demand for this text, closely related to practices of the imperial horse management policy and the movement of horse around the Kingdom in connection to the military and Imperial postal system in certain regions of the Ming Dynasty.

Horse Administrative Policies, and the Imperial Postal System in the Ming Dynasty

The imperial court of the Ming Dynasty devoted distinguished attention to the so-called *ma zheng* "horse policy", as it adopted and developed the imperial horse administration system developed in the previous Mongol Yuan Dynasty (1271–1368), including the imperial postal network as well as the imperial horse administration system, all requiring healthy horses that could be reliably ridden long distances.

Horses in Warfare

Having driven the Mongols out of the North China Plain, the Ming Dynasty still faced severe military threats from nomadic powers along its northern border. Setting up a tightly organized imperial horse administration system ensured a constant supply of military horses ensuring military mobility along the border. This mobility became increasingly necessary during the early Ming period, when the Ming court still applied an aggressive military policy against nomadic horsemen raiding from the north (He Liping 2005, 101). The founder of the Ming Dynasty, Emperor Taizu (ruling between 1368 and 1398), especially emphasized the importance of maintaining an effective horse administration. He announced that the number of available horses was the main criterion for qualifying a country as wealthy and powerful (Hu Shi'e ed. 1991, 70). It is also during his reign that the Ming court formulated a regularized administrative system dealing with horse administration. In 1402, his successor, Emperor

Chenzu (ruling between 1402 and 1424), questioned the minister of the *bingbu shangshu* "Ministry of War" about the number of available horses. He replied that there were only around 23,700 registered horses available for the government to use at that moment because earlier wars against the nomadic powers consumed a significant number of horses (Yu Jideng ed. 1995, 268). Facing this strategic challenge, Emperor Chenzu continued regulating the system of imperial horse administration, through practicing more systematic and detailed policies of horse management.

Laws and decrees regulated the Ming Dynasty's imperial horse policy in great detail. According to the chapter devoted to horse administration in the dynastic chronicle of the Ming Dynasty, there were four major types of institutes in governmental organization. They were responsible for both governmental and private horse breeding in different regions (Table 3.2).

Horse breeding under the supervision of the Ming imperial government consisted of two regime types. The first was the so-called *guan mu* "official horse breeding", referring to keeping horses on a state-owned pasture by the regional army. The second was *si mu* "private horse breeding", which referred to a mandatory policy that assigned local peasants, people in the so-called *ma hu* "horse household", to keep the state's horses on their own farms and pastures. The horse households were obliged to regularly deliver a certain number of horses to local supervisory authorities. This obligation was closely controlled and any failure or fault was severely punished. The mandatory horse household system played a crucial role in Ming horse administration, providing considerable numbers of military horses for the imperial armies stationed at the capital cities, namely present-day Beijing and Nanjing. The imperial government acquired unquestioned ownership of the horses within this dual-tier system. Meanwhile, the communities obliged to keep and manage the imperial horse stock varied greatly between different regions.

The Ming Dynasty Postal Network

Besides the imperial horse breeding system designed for maintaining a regular supply of military horses, the Ming Dynasty also needed horses for the extended imperial postal network developed by the Mongol Yuan Dynasty and taken over by the Ming rulers. This network was based on thousands of postal relay stations, requiring a huge number of healthy horses, as one of the most important animal transportation tools.

Table 3.2 Ming governmental institutes responsible for horse administration (horse policy "*ma zheng*" in *Ming shi*, 2270–2277)

Institute	*Location*	*Scope of official duty*	*Type of horse keeping*	*Supply target*
Yuma jian "Directorate of Imperial Horses"	Beijing	supervising horse breeding at the imperial court	official	Imperial court
Taipu si "Imperial Court Stud"	Beijing/Nanjing	supervising horse breeding in the North and East China Plains	non-governmental	Imperial army stationed in capitals
Xing Taipu Si "Moving Branch of the Imperial Court Stud"	Liaodong/Beijing/Shanxi/Gansu/Shannxi	supervising horse breeding at frontier military fortresses	official	Frontier army
Yuanma Si "Pasturage Office"	Beijing/Liaodong/Gansu/Shannxi	supervising horse breeding on regional imperial pastures	official	Frontier army

Recently, comprehensive research has focused on the imperial postal system in Ming China (Kauz 2005, 75–89; Harris 2015, 287–310; Zi Wang 2017, 59–106). As mentioned previously, this institution was another administrative heritage from the preceding Mongol Yuan Dynasty. According to military treatises in the dynastic chronicle of the Yuan Dynasty (*Yuan shi*) edited by the Ming court, an extensive postal relay system was continued into the early Ming period. It was based on *zhan chi* "postal relay stations" and *jidi pu* "military express couriers", elements in a communication system formulated by the Yuan court during the late thirteen century. Another partially preserved legal text written around 1330, *jingshi dadian* "compilation of administrative documents", contains a detailed register listing the locations and scale of around 1350 postal stations in Yuan Dynasty China. It is possible that this number is an underestimate as numerous names of other stations can also be found in other primary sources (Olbricht 1954, 24). During Yuan rule, the main function of imperial postal stations was to supply food as well as means of transportation such as horses, oxen, and carts for privileged officials and aristocracy during long-distance travel. The fourteenth century Chinese register on the administration of postal stations in Heichen fortress (present-day Halahaote, Inner Mongolia), recorded the detailed types and quantities of food, the so-called *zhiying* "humble offering", in contemporary texts, provided by local postal stations (Chen Gaohua 2002, 49–56; Ye Xinmin 2007, 86–93).

The Ming Dynasty almost completely adopted the Yuan administration's policy concerning its postal system. The Ming imperial postal relay system likewise consisted of three basic units, which were respectively the *yi zhan* "postal relay station", mostly for dispatching envoys, the *jidi pu* "express couriers" for delivering official documents, and the *diyun suo* "transport station", chiefly for delivering military goods (Yang Zhengtai 2006, 4). The imperial postal system, established some time in the early Ming period, was originally exclusively designed for official and military purposes at the time, as shown by coeval legal documents recorded in *Da Ming Huidian* "compilation of the Great Ming's laws" edited in 1587. This document contains detailed entries that strictly regulate the usage of imperial postal services to a limited cadre of qualified officials and aristocratic people. An ordinance published in 1393 by Emperor Taizu, stated that only that set of official envoys and military messengers were qualified to use the imperial postal relay services (Sheng Shixing et al., eds 1989, 2937). However, by the middle Ming period, irregularities in using the imperial postal system became increasingly common. Local officers supervising the administration of local postal stations, abused their privileges to use the postal system, becoming engaged in private and illegal trade. Even low-level servants at postal stations could acquire illegal profits through participating in illegal trade and selling permissions to unqualified people. Hu Zuanzong, an official-intellectual active around the fifteenth century, complained that, "over 80 percent of today's postal activities were exploited illegally while legal activities comprised less than 30 percent" (Hu Zuanzong 1995, 431). Another contemporaneous intellectual-official named Zhang Xuan (ca 1553 to 1636) also recorded that "ranging from (high-ranking) officers in capital cities to local officers within counties and (low-level) servants, everybody abuses his power to illegally benefit from the postal system, through either carrying out illegal private trade or reselling permissions" (Zhang Xuan 1980, 215). Although by Ming law, illegal use of the postal system carried a death penalty in severe cases, the law became very difficult to enforce as the black-market activities proved extremely lucrative. A selected collection of legal documents and juridical cases printed in 1578, the *Xianzhang leibian* "Categorized collection of legal documents", recorded that even an administrator of a postal station, one of the lowest-level official positions in the Ming bureaucratic hierarchy, could gain "thousands *liang* (tael) of silver ingots" per year through accepting merchants' bribes and pursuing illegal trade (Su Tongbing 1969, 200). These records show that during the middle Ming period, the imperial postal system was only nominally used for official missions, since it was commonly exploited illegally by local merchants and corrupt officials for commercial or other private purposes. Beyond being illegal, such abuses show that due to its convenience and efficiency, the imperial postal system also played an increasing, albeit unofficial, role in interregional trade and communication in China at that time. Consequently, the excessive exploitation of the postal network beyond its legal capacity created an increased demand for horses as means of transportation in supporting the operation of most local postal stations. Reliance on horses in particular, increased in regions that had relatively well-developed local market economies.

Horses and Taxation

As mentioned before, one of the major purposes of horse administration policy in Ming Period China was to ensure a sufficient supply of horses for the imperial postal system. In the early Ming period, under the reign of Emperor Taizu, a number of wealthy households from the rich and prosperous Yangtze Delta region were forced to purchase horses for the state, by order of the emperor (Zhang Tingyu et al., eds 1974, 1845). Later on, this policy was transformed into a *de facto* mandatory taxation system levied on registered local households in wealthy cities in the Yangtze Delta, lasting until the seventeenth century according to the writings of Mao Yuanyi (1594–1640), a prominent military scholar of the age (Mao Yuanyi 1997, 378). Besides such monetized taxation levied on the local population for purchasing and keeping the horses, a certain number of local rural households, especially in the regions neighboring the two imperial capitals of the Ming Dynasty, were directly assigned to permanently keep horses for the state for generations. In 1373, Emperor Taizu decreed that every eleven registered households in such horse-breeding regions in South China had to keep one horse, and the government would supply its fine horses for breeding. These assigned horse-breeding households had to submit a foal every year and were punished by forced labor if they failed to comply with the imperial assignment. During the reign of Emperor Chenzu, this policy was slightly modified due to to the growth of the human population. The number of horses assigned was no longer based on the number of households but on the number of people in the population. In 1415, Emperor Chenzu decreed that every 15 registered adult males should provide a foal annually. Failure to complete this assigned quota or loss of horses had to be financially compensated for by the horse-breeding households. The mandatory horse breeding policy based on human population size lasted for over two centuries in the Yangtze Delta region, until the end of the Ming Dynasty.

YHLMJ and the Ming Dynasty's Horse Administration Policy in the Nanzhili Region

The Yangtze Delta region, largely corresponding to the Nanzhili administrative district, supplied the Ming Dynasty's southern capital Nanjing. This region also played a pivotal role in both the aforementioned imperial postal network and imperial horse breeding system.

Horse Populations in Nanzhili

Already by the era of Emperor Taizu, Nanzhili had become one of the major horse-breeding regions in the Ming Dynasty. It is also in this region that local authors and private publishers produced several types of veterinary texts on animal disease, especially during the late sixteenth and early seventeenth century. These works include YHLMJ. However, the appearance of YHLMJ is not simply a direct consequence of the practices of horse policy in the Nanzhili region, as its emergence and spread were related to certain changes in horse policy and even the natural environment in this period.

The distribution of imperial horses among county-level districts in Nanzhili was primarily based on the number of registered local residents (Fig. 3.2). The Court of the Imperial Stud at Nanjing's imperial court (*Nanjing taipusi zhi*, hereafter TPSZ; Lei Li ed. 1996) was the institution responsible for administrating the local horse households. According to its records, in principle, every 10 registered adult men in the Nanzhili region

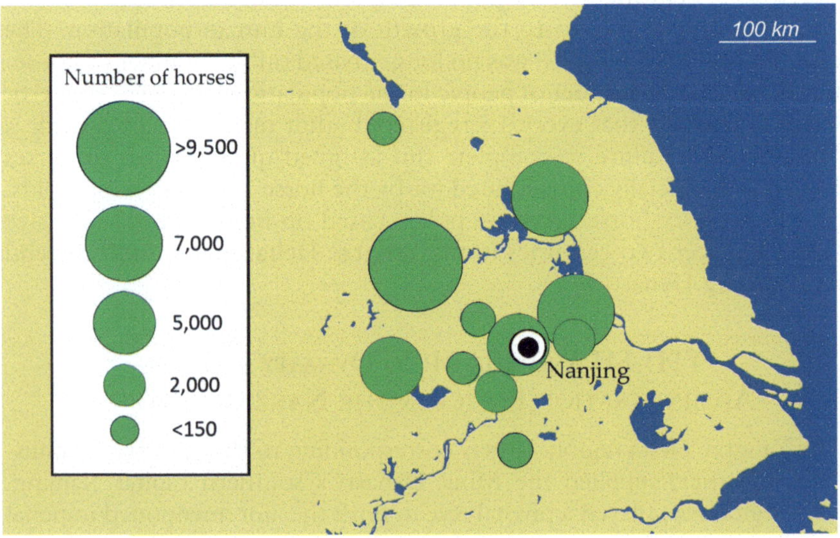

Fig. 3.2 The number of imperial horses kept in local pastures within the Nanjing Metropolitan area (*Nan Zhili*) around 1550. (Source: *Nanjing Taipusi zhi*. Original drawing)

were assigned one stallion, and every 15 took care of one mare (TPSZ j.10, 1). These quotas were theoretical numbers defined some time during the early Ming period. In practice, especially during the late Ming period, the actual ratio of imperial horses kept by local inhabitants was much lower due to the heavy burden of this assignment and associated taxation, concomitant corruption as well as human population growth.

By the last quarter of the sixteenth century, Nanzhili was the most populated province-level district in Ming Dynasty China, a place where 17.3 percent of its known population (10.5 million people) were registered (Sheng Shixing et al., eds 1989, 123). However, in a largely coeval sixteen-volume official gazetteer edited around 1550 on the TPSZ, over 30 percent of imperial horses of the Ming Dynasty were kept in the Nanzhili area, comprising around 37,500 horses (TPSZ, 1). This suggests that while, in practice, the imperial horse policies did not achieve their goal planned in the aforementioned legal text, Nanzhili was still the most important region for imperial horse breeding, given the proportion between its local human population to imperial horses. In addition, it implies that the large scale of horse breeding in Nanzhili was not simply a result of its sizable human population, but also a consequence of intentional Ming court policies. Since the early Ming period, laws and regulations were intended to invest significant horse breeding programs in the regions surrounding imperial capitals.

The Postal Network in Numbers

In addition to the previously discussed high concentration of horses, an extensive and densely arranged imperial postal network based on routes radiating from Nanjing, strengthened the connection between local society and horses. A 1570 description of postal stations and related routes of the time was compiled by a local merchant, Huang Bian, who lived in Xiuning County of Nanzhili (present-day Anhui Province, East China). Huang Bian was an experienced long-distance commercial traveler (Huang Bian ed. 1994). It was possible to draft a basic map illustrating postal stations and routes using his writings (Fig. 3.3).

In 1570, Huang Bian's *Yitong Lucheng Tuji* "an illustrated list of postal routes" mentioned ten main travel routes connecting Nanjing with other provinces in detail. There were approximately 94 postal stations along these routes within the Nanzhili region. Over 80 of them were overland stations supporting transport. They chiefly relied on horses and oxen as

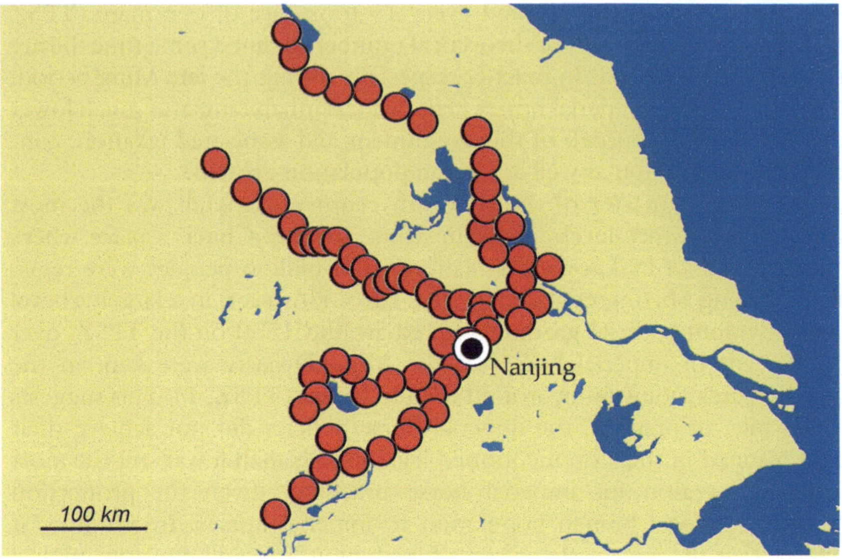

Fig. 3.3 Postal stations within Nanzhili recorded in *Yitong Lucheng Tuji* (1570). (Source: Huang Bian ed. 1994. Original drawing)

their transportation tools. According to my analysis, the average distance between each postal station was around 46 *li* (ca 24.8 km) in Nanzhili, which is slightly shorter than the theoretical distance of about 60 to 80 *li* recorded in the 1368 decree issued by Emperor Taizu on the establishment of the postal stations (Xia Yuanji et al., eds 1962, 500). The 46 *li* distance is also noticeably shorter in comparison with corresponding data from other regions during the same period. The average distance of postal stations in other provinces in northern China of the Ming Dynasty normally ranged between 30 and 40 km (Sun Xifang 2010, 27–35; Wang Mianhou and Piao Wenying 2016, 438–457) and was closer to the theoretical value of 60 to 80 *li* recommended in 1368. To some extent, the relatively hilly and marshy terrain of the Nanzhili region must have played a role in developing this more densely arranged postal network, but one of the fundamental factors must still have been the density of settlements as well as that of their human population.

Regulating Horse Households

In order to further contextualize the appearance and prevalence of the veterinary text YHLMJ within its contemporary social environment, it is necessary to look at the aforementioned activities related to the imperial horse policy from a more grass-roots perspective. Scattered descriptions among writings of this period, still give researchers a chance to illustrate the living conditions within these horse households. According to laws and legal documents of the time, horse households and basic level bureaucrats who failed to submit or deliver the required horses by the appointed time were severely punished. For instance, if the horses provided by the government for local households to breed accidently died, then the horse households had to pay a fine to compensate for the loss (TPSZ 1, 1).

According to a 1466 decree, if basic level officers failed to pass their supervising institutes' annual examination of their work concerning the promotion of horse breeding, they even faced corporal punishment in the form of public flogging (He Qiaoyuan 1997, 135). Yang Shiqiao (1531–1609), director of the Court of Imperial Stud in Nanjing around the late sixteenth century, mentioned in his *mashu* "treatises on horse policy" printed in 1594 that the horse breeding assignment had become a heavy burden for lower class people, as even people with physical disabilities and widows, who were normally exempt from such labor, would be forced to share the cost of horse breeding in rural areas (Yang Shiqiao 1984, 143). The main motivation behind the Ming imperial court developing mandatory horse breeding by peasants in Nanzhili region was the proximity of Nanjing. People living around imperial capitals were obliged to keep horses according the Ming law that had introduced the institution of horse households in the first place (TPSZ j.1, 1.).

However, competition for the land, including pasture lands, by the ever-increasing human population in this densely inhabited region intensified, especially after the middle Ming period. There are academic controversies over the exact quantity of land available for horse-keeping and farming activities during the late Ming period. It is usually accepted however, that the *per capita* arable area was the lowest in Nanzhili among all Ming province-level districts (Table 3.3). The sixteenth-century gazetteer TPSZ recorded the numbers and areas of registered local pastures in Nanzhili under the administration of the Court of Imperial Stud at Nanjing as projected on the map in Fig. 3.4. The resulting pattern suggests that vast areas in the densely populated prefectures and counties

Table 3.3 Demographic data and the number of registered horses in Nanzhili's prefectures (Source: TPSZ). One *qin* equalled ca 56,655 m² during the Ming era

Prefecture	*Registered adult men*	*Horses total*	*Mare*	*Stallion*	*Number of pastures*	*Area of pastures (qin)*
Yingtianfu	66,777	4666	3728	932	442	1445
Zhenjiangfu	36,210	2340	1872	468	65	151
Taipingfu	18,510	1463	1172	293	69	622
Nanlin	7800	750	600	150		
Jianping	11,200	800	640	160	3	26
Luzhoufu		4374	3499	875	134	670
Chuzhou		1075	860	215	42	130
Hezhou		637	510	127	83	395
Fengyangfu	14,340	9476	7578	1898	130	480
Xuzhou		150			28	12
Huai'anfu	82,000	6310	5048	1262	104	792
Yangzhou	36,900	4743	3794	946	260	1367

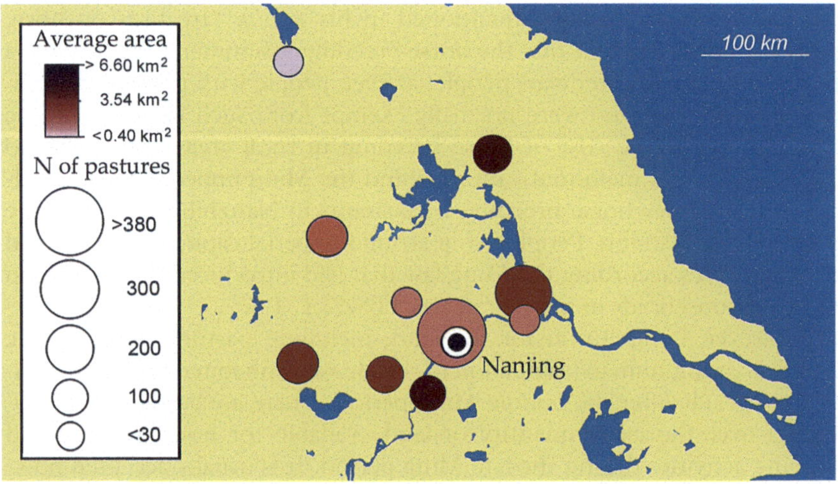

Fig. 3.4 The quantity and average area of local pastures in Nanzhili's prefectures under the supervision of *Taipu si* (Court of Imperial Stud) at Nanjing, around 1550. (Source: *Nanjing Taipusi zhi*. Original drawing)

along the Yangtze and Huaihe rivers were used as horse pastures. This governmental arrangement further intensified the competition for land in the Nanzhili region. In this situation, the primary concern of horse households as well as responsible local officers was to keep the horses healthy and productive. Otherwise, the financial consequences would have been ill-afforded. Many local households, unable to sustain horse breeding fled to other regions to escape punishments such as forced labor (Yin Luanzhang 1984, 451). In 1567, a report written by the Vice Director in the Ministry of Personnel (*libu shilang*) to Emperor Muzong (ruling between 1567 and 1572) mentioned that many horse households in Nanzhili had fled to avoid having to keep horses as low levels of grain production had become insufficient for human survival (Huang Yunmei 1984, 853).

Maintaining the Ming infrastructure evidently required huge numbers of horses, creating a demand that was difficult to meet. Policy changes aimed at increasing the horse supply imposed heavily on the local economies in the Nanzhili region where people were forced to provide horse care for the state. After the middle of the sixteenth century, this changing situation created a need for pragmatic veterinary manuals instructing producers on horse breeding issues. As professional state-appointed veterinarians working in local pastures at Luzhoufu, Nanzhili, the Yu brothers were quite familiar with the mandatory imperial horse breeding system practiced in Nanzhli, since it was one of their assigned duties to seasonally examine the horses submitted by low-level officers as well as horse households.

Concerning its content, YHLMJ is not simply a veterinary text, but also a useful manual intentionally aimed at instructing local horse households on how to keep the state's horses. One of the main points of YHLMJ concerns successful foaling. Consequently, the entire fourth volume (consisting of five treatises on pregnant mares), is chiefly dedicated to this theme. In addition, other volumes also contain entries related to the mandatory horse breeding policy. For instance, in the fourth volume (*juan*), there are six entries arguing for the importance of keeping the horses healthy, especially in December and March. On the one hand, the often capricious weather in these months might easily cause certain horse diseases. On the other hand, these months were important for the official examination procedure aimed at maintaining the quality of horses. According to *Mazhengji*, local officers should calculate the number and examine the state of pregnant mares by December. In March, the required number of horses were sent to an appointed institution for physical examination. Qualified horses then had to be submitted to the imperial stud farms. Evaluations of horse households and local officers were based on their results.

Entries on veterinary remedies in the second volume discuss herbal ingredients applied by the authors. According to their own introduction, most of these herbs were indigenous to the Nanzhili region. This, in part, shows their firm knowledge and experience in applying local ingredients, while it also implies that their target audiences were horse households living in the Nanzhili region where these remedies were locally available.

Another noteworthy change in the conditions of horse breeding during the time of YHLMJ, the fifteenth and sixteenth centuries, was the frequent occurrence of infectious horse diseases.

There are a number of studies focusing on the impact of environmental factors on the prevalence of human epidemics in the middle and late Ming period. Coeval infectious animal diseases are much less discussed (Chen Xu 2016).

Climate Change

From a global perspective, the concept of "Little Ice Age" (ranging largely from the early thirteenth century to the middle nineteenth century), is closely associated with the occurrence of climate-related historical events. These include the Great Famine, the Black Death, and the Thirty Years' War in Europe (Hoffmann 2014, 319). In Central Asia, some authors (Pederson et al. 2014; Putnam et al. 2016) link the Little Ice Age with the expansion of grasslands to cooling lower latitudes that improved conditions for equestrian lifestyles, politically strengthening the Mongol Empire.

Recently the rise and decline in the dynastic cycle from the Yuan to the Ming Dynasty has also attracted considerable attention. Certain scholars suggest that deadly infectious diseases and natural disasters connected to climate change weakened imperial control over regional societies. This presumably facilitated large-scale rebellions that eventually lead to the overthrow of central governance (Brook 2013, 106–133; Anderson 2015, 12–29).

Meanwhile, regional responses to major shifts in climate such as the Little Ice Age would have been widely variable across Eurasia or even within the larger Ming Empire. The cooling that possibly improved desert conditions in the north-east may have had adverse effects in the subtropical south. Increasing humidity may have caused floods resulting in crop failure and deteriorating water sanitation (Shuman 2010, 1062).

According to Chen Xu's statistic concerning the frequency of severe infectious human diseases during the Ming Dynasty, Nanzhili suffered most often. Historians know of 25 severe epidemics that struck this region during this period (Chen Xu 2016, 13). Additional research on the frequency of natural disasters such as floods, famine, and storms in southern China during the Ming Dynasty, also shows that most were concentrated in the late sixteenth and early seventeenth century (Wang Shuanghuai 1999, 157). While there is not always a direct causal relationship between the occurrence of human epidemics and animal infectious diseases, an unstable environment frequently hit by natural disasters may have been a common background factor promoting both.

In YHLMJ, the author applied a popular belief about infectious diseases widely held at the time: epidemics or plagues were seen as the consequence of the intrusion of the so-called *yiqi* "air of polluted external environment" into the body. In the second volume of YHLMJ, 30 types of horse diseases are categorized as maladies caused by external noxious factors, including the air, the land, and the water. For instance, under the entry of eye diseases in horses, the authors indicate that some are caused by the intrusion of *yiqi* in the liver of the horses caused by a noxious external environment. In the treatise discussing internal diseases in horses, the appearance of such noxious factors is due to an imbalance between Earth and Heaven. This theory can also be found in texts about human disease written during the sixteenth century (Xu Zhexin 2017). In the first volume of YHLMJ, there are four treatises about sanitation in horse stables, stating that it is important to protect horses from noxious external factors, and it is necessary to separate sick horses from healthy ones (Yu Benyuan and Yu Benheng ed. 1957, 19–21). This implies that Chinese veterinarians of the time must have been aware of the infectious nature of certain horse diseases. Perceptions of horse diseases reflected in YHLMJ are closely related to the understanding of epidemic diseases by Chinese medical scholars of the period. The prevalence of plagues and frequent natural disasters in the sixteenth century presumably influenced the perceptions of veterinarians in Ming period China, as is apparent in certain cases in YHLMJ. However, targeted research is needed to answer the question of whether environmental changes really caused increasing animal morbidity, and whether this phenomenon stimulated the production of veterinary texts in the late Ming period in China.

Conclusions

As successor to the Mongol Yuan Empire, the Ming Dynasty (1368–1644) relied heavily on horses, not only as a means of military control, but also as part of the continuing development of an extensive postal system administratively linking key areas in China. In order to support such a large-scale postal network, the demand for healthy horses as means of transportation was already increasing in the early Ming period, especially in mountainous southern China where local horse husbandry was less developed than in the plains of the north.

In an effort to increase the supply of horses, the Ming government imported horses from neighboring regions in Central Asia, Tibet, and Ryukyu through commerce and tributary trade. In addition, the government also forced local peasant households to keep and breed horses. Low-ranking officials at postal stations were compelled to breed horses as well. Within this historical context, knowledge about equine breeding and medicine was of key importance, especially in regions where horse husbandry traditions had not been developed previously. Through analyzing the authorship, content, audience, and influences of YHLM within this context, a close connection can be seen between the rise of the Ming period imperial postal system and the publication of equine veterinary texts. The horse transited from a simple means of production to an integral part of imperial administration and became a driving topic in coeval scholarly inquiry. In addition, this text was not only a product of developing horse husbandry in agricultural Chinese society, but also a work illustrating cultural and knowledge transfer between China and its neighboring regions of Mongolia, Korea, and Japan.

References

Anderson, Mary. "The not so little Little Ice Age: Climate Change Impacts on the Collapse of the Yuan and Ming Dynasty." *Emory Endeavors Journal* 6 (2015). 12–29.

Brook, Timothy. "Chapter 5: Economy and Ecology." In *The Troubled Empire. China in the Yuan and Ming Dynasties*, edited by Timothy Brook, 106–133. Cambridge MA: Harvard University Press, 2013.

Buell, Paul D., Timothy May, and David Ramey. "Chinese Horse Medicine: Texts and Illustrations." In *Imagining Chinese Medicine*, edited by Vivienne Lo, and Penelope Barrett, 315–326. Sir Henry Wellcome Asian Series, vol. 18. Leiden: Brill, 2018.

Chen Gaohua. “Heicheng Yuandai zhanchi dengjibu chutan” [A study of Heichen register documents about Yuan China’s postal stations]. *Zhongguo shehui kexueyuan yanjiusheng xuebao* 5 (2002). 49–56.

Chen Xu. 2016. *Mingdai wenyi yu Mingdai shehui* [Epidemic disease and local societies in Ming China]. Chengdu: xi’nan caijing daxue chubanshe.

Cooke, Bill. 2000. *Imperial China: The Art of the Horse in Chinese History: Exhibition Catalog*. Kentucky Horse Park. Lexington KY: Art Media Resources Limited.

Feng Hongqian. “Ming yuanheng liaomaji xiancun banben kao” [A bibliography of currently preserved editions of yuanheng liaomaji]. *zhong shouyixue zazhi* 2 (2008). 29–30.

Harris, Lane J. “The ‘Arteries and Veins’ of the Imperial Body: The Nature of the Relay and Post Station Systems in the Ming Dynasty, 1368–1644.” *Journal of Early Modern History*, 19 no. 4 (2015). 287–310.

He Liping. “Luelun Mingdai mazheng shuaibai ji dui guofang yingxiang” [A study of the decline of Ming China’s horse administration system and its influences on national defense]. *Junshi Lishi Yanjiu* 1 (2005). 98–103.

He Qiaoyuan. “Mazhengji” [Notes on horse policy]. In *Siku jinhuishu congkan shibu* [Compilation of prohibited books in the Siku collection: history category]. Vol. 47, 135–144. Beijing: Beijing chubanshe, 1997.

Hoffmann, Richard C. 2014. *An Environmental History of Medieval Europe*. Cambridge: Cambridge University Press.

Hu Shi’e ed. 1991. *Ming Taizu Ji* [Collection of Decrees and Ordinances issued by the Taizu Emperor of the Ming Dynasty] Hefei: Huangshan shushe.

Hu Zuanzong. 1995. *Yuanxue Bian* [Essay collection on the motivation to study], vol. 2. Ji’nan: qilu shushe.

Huang Bian. “Yitong Lucheng Tuji” [An Illustrated List of Postal Routes]. In *Mingdai yizhan kao* [A Study of Ming China’s Postal Stations], edited by Yang Zhengtai, 215–219. Shanghai: Shanghai guji chubanshe, 1994.

Huang Yunmei. 1984. *Mingshi kaozheng* [Study on Ming history], vol. 3. Beijing: zhonghua shuju.

Kauz, Ralph. “Postal Stations (Yizhan 驛站) in Ming China (1368–1644).” In *Trade and Transfer Across the East Asian ‘Mediterranean’*, edited by Angela Schottenhammer, 75–89. Wiesbaden: Harrassowitz, 2005.

Lei Li ed. “Nanjing Taipusi zhi” [Gazetteer of the Court of Imperial Stud at Nanjing]. In *Siku quanshu cunmu congshu shibu* [Compilation of preserved categorized books in Siku collection: category of history]. Vol. 257, 487–623. Ji’nan: Qilu shushe, 1996.

Mao Yuanyi. 1997. “Zhangji” [Sketches on palms]. In *Siku jinhuishu congkan jibu* [Compilation of prohibited books in the Siku collection: category of essay collection]. Vol. 110, 359–398. Beijing: Beijing chubanshe, 1997.

Meserve, Ruth I. "Chinese Hippology and Hippiatry: Government Bureaucracy and Inner Asian Influence." *Zeitschrift der Deutschen Morgenländischen Gesellschaft* 148 (1998). 277–314.

Olbricht, Peter. 1954. *Das Postwesen in China unter der Mongolenherrschaft im 13. und 14. Jahrhundert.* Wiesbaden: O. Harrassowitz.

Pederson, Neil, Amy E. Hessl, Nachin Baatarbileg, Kevin J. Anchukaitis, and Nicola Di Cosmo. "Pluvials, droughts, the Mongol Empire, and modern Mongolia." *Proceedings of the National Academy of Sciences* 111, no. 12 (2014). 4375–4379.

Putnam, Aaron E., David E. Putnam, Laia Andreu-Hayles, Edward R. Cook, Palmer, Jonathan G., Elizabeth H. Clark, Wang Chunzeng, Chen Feng, George H. Denton, Douglas P. Boyle, Scott D. Bassett, Sean D. Birkel, Javier Martin-Fernandez, Irka Hajdas, John Southon, Christopher B. Garner, Cheng Hai, and Wallace S. Broecker. "Little Ice Age wetting of interior Asian deserts and the rise of the Mongol Empire." *Quaternary Science Reviews* 131, Part A (2016). 33–50.

Ramey, David W., and Bernard E. Rollin. 2003. *Complementary and Alternative Veterinary Medicine Considered.* Ames IA: Iowa State Press.

Robinson, David. "Why military institutions matter for Ming history." *Journal of Chinese History* 1 (2017). 297–327.

Sheng Shixing et al., eds. 1989. *Ming huidian* [Compilation of Ming's Laws]. Beijing: zhonghua shuju.

Shuman, Emily K. "Global Climate Change and Infectious Diseases." *The New England Journal of Medicine.* March (2010). 1061–1063.

Su Tongbing. 1969. *Mingdai yidi zhidu* [The postal relay system in Ming China]. Taipei: zhonghua congshu bianshen weiyuanhui.

Sun Xifang. "Mingdai shanbei diqu yizhan jiaotong de fazhan jiqi dui junshi jingji de yingxiang" [The development of postal stations and its impact in Ming China's Shanbei region]. *Chang'an daxue xuebao* 4 (2010). 27–35.

Wang, Chelsea Zi (2017). Zi Wang, Chelsea Zi. "*Dilemmas of Empire: Movement, Communication, and Information Management in Ming China, 1368–1644*". PhD diss, Columbia University NY, 2017.

Wang Mianhou, and Wenying Piao. 2016. *Zhongguo dongbei yu dongbeiya gudai jiaotongshi* [The history of transportation and communication in ancient Northeast China and Northeast Asia]. Shenyang: Liaoning renmin chubanshe.

Wang Shuanghuai. "Mingdai Hua'nan de ziran zaihai jiqi shikong tezheng" [A discussion on the natural disasters in South China of the Ming Dynasty]. *Dili Yanjiu* 18 (1999). 157.

Xia Yuanji et al., eds. 1962. *Ming Taizu shilu* [Chronicle of the Taizu Emperor]. Vol. 29. Taipei: Academia Sinica.

Xu Zhexin. "The Environment, Perceptions, and Publication of Medical Texts in Fujian during the Ming Period (1368 to 1644)." Paper presented at the conference *Seafaring, Trade, and Knowledge Transfer: Maritime Politics and Commerce in Middle Period and Early Modern China*. Ghent: Ghent University, Het Pand, 2017.

Yang Jialuo ed. 1975. *Ming shi* [Chronicle of the Ming Dyansty]. Taipei: dingwen shuju.

Yang Shiqiao. 1984. *Xinke mashu* [Treatises on horse policy]. Beijing: nongye chubanshe.

Yang Zhengtai. 2006. *Mingdai yizhan kao* [A study of Ming China's postal stations]. Shanghai: Shanghai guji chubanshe.

Ye Xinmin. "Yuandai yizhan de zhiying zhidu" [The food supply and services provided by Yuan China's postal stations]. *Mengushi yanjiu* 9 (2007). 86–93.

Yin Luanzhang. 1984. *Mingjian* [History of Ming]. Shanghai: Shanghai shudian chubanshe.

Yu Benyuan, and Benheng Yu. 1736. *Yuanheng liaomaji*. 437.4/432(1), Library of Nankai Univeristy.

Yu Benyuan, and Benheng Yu. 1957. *Yuanheng liaomaji*. Beijing: nongye chubanshe.

Yu Jideng ed. 1995. *Huang Ming Diangu Jiwen* [Literally Allusions and Anecdotes of the Ming Dynasty]. Beijing: shumu wenxian chubanshe.

Yu Yuan. 1573. *Tuxiang shuihuangniu jing hebing daquan* [Compilation of illustrated treatises on the diseases of oxen] Nanjing: Shulin jubao tang. AA.47, Municipal Library of Yonezawa, Japan.

Zhang Xuan. "Xiyuan jianwenlu" [Things seen and heard by mister Xiyuan]. In *Mingdai shehui jingji shiliao xuanbian* [Selected historical source book of Ming China's economy and society] vol. 3, edited by Guozhen Xie, 214–218. Fuzhou: Fujian renmin chubanshe, 1980.

Zhang Tingyu et al., eds. 1974. *Ming shi* [Dynastic chronicle of the Ming Dynasty]. Beijing: zhonghua shuju.

CHAPTER 4

Medieval Animals: The Fast and the Slow

Gerhard Jaritz

... nos deux mondes se rejoignent.
L'animal peut faire parfois mieux que l'homme.
Il peut également faire aussi mal.
(Lenoir 2017, 81)

The interest in animals in the Middle Ages has several explanations: economic perspectives, social aspects, agricultural, alimentary, medical, religious, or "natural scientific" viewpoints, and so on. All of these interests and perspectives were certainly not neutral but have to be seen as connected with particular positive or negative qualities and evaluations of specific animals, as well as with the products based on or assembled out of them. One of them, with regard to the living animals themselves, is certainly their respective speed or slowness. One regularly finds references to those characteristics in a variety of textual and visual sources and source types, and in different contexts. They are often connected with comparisons between specific animals or between animals and humans in an allegorical or metaphorical way. This contribution aims at offering a sample of

G. Jaritz (✉)
Department of Medieval Studies, Central European University, Vienna, Austria
e-mail: jaritzg@ceu.edu

L. Bartosiewicz, A. M. Choyke (eds.), *Medieval Animals on the Move*, https://doi.org/10.1007/978-3-030-63888-7_4

such fast and slow animals and the variety of their contextualizations, mainly based on medieval "natural histories," as well as encyclopedic and religious literature which are, now and then, also complemented by visual representations of the described animals.

In natural historical treatises there is a regularly mentioned link between the names of animals with their qualities that also sometimes refers to velocity or slowness, often based on Isidore of Seville's (ca 560–636) *Etymologiae* that, in general, became a model for many later authors. The hare, *lepus*, for example, received its name from *levipes* (light foot), because it runs fast: *Lepus, levipes, quia velociter currit. Vnde et Graece pro cursu λαγώς dicitur; velox est enim animal et satis timidum* (Barney et al. 2006, 248, XII.I.23; Lindsay ed. 1971, XII.I.23). The dromedary is speedy. It received its name from the Greek ϑρόμος, meaning running and speed: *Dromeda genus camelorum est, minoris quidem staturae, sed velocioris. Vnde et nomen habet; nam δρόμος Graece cursus et velocitas appellatur* (Barney et al. 2006, 249, XII.I.36; Lindsay ed. 1971, XII.I.36). The fox is named *vulpis*, as it is quick on its feet (*volubilis pedibus*): *Vulpes dicta, quasi volupes. Est enim volubilis pedibus, et numquam rectis itineribus, sed tortuosis anfractibus currit, fraudulentum animal insidiisque decipiens* (Barney et al. 2006, 253, XII.II.29; Lindsay ed. 1971, XII.II.29). The vulture (*vultur*) is said to have gotten its name from its slow flight (*volatu tardus*), the result of having a large body: *Vultur a volatu tardo nominata putatur: magnitudine quippe corporis praepetes volatus non habet* (Barney et al. 2006, 264, XII.VII.12; Lindsay ed. 1971, XII.VII.12).

Isidore influenced many later authors by stating that the tiger (*tigris*) got its name from the word the Persians and Medes used for "arrow" because tigers are so fast. Likewise, the Tigris River is called by this name because it is the fastest of all rivers:

> *Tigris … est enim bestia … velocitate mirabilis; ex cuius nomine flumen Tigris appellatur, quod is rapidissimus sit omnium fluviorum.* (Barney et al. 2006, 251, XII.II.7; Lindsay ed. 1971, XII.II.7; Boese ed. 1973, 163–64)
>
> *Tigris fluvius Mesopotamiae de Paradiso exoriens et pergens contra Assyrios, et post multos circuitus in mare Mortuum, influens. Vocatus autem hoc nomine propter velocitatem, instar bestiae tigris nimia pernicitate currentis.* (Barney et al. 2006, 281, XII.XXI.9; Lindsay ed. 1971, XII.XXI.9)

The velocity of the tiger could also lead to a phenomenon mentioned by Thomas of Cantimpré (1201–1272) in his *Liber de natura rerum*. In the description of the European bison (*De zuprone*), he also refers to its particular velocity. This led, as he says, to a situation where people gave it the name tiger. However, they are wrong, as the very fast tiger lives in the East and the bison is found in Bohemia:

> *De zubrone. Zubrones bestie ferocissime sunt et sunt de genere taurorum silvestrium. ... Summae celeritatis sunt; falsoque obtinuerunt in illis regionibus a Latinis nomen tygris, et hoc propter velocitatem. Errant enim qui hoc dicunt, quia tygris bestia satis modica et velocissima est orientem inhabitans. Hec vero in Boemia reperitur.* (Boese ed. 1973, 172)

Sometimes, the textual treatment of quick or slow animals is limited to simple descriptions or positive or negative statements with regard to their velocity or slowness, as found, for instance, in a passage concerning dogs in Isidore's *Etymologiae* stating that they have two qualities: bravery and speed: *In canibus duo sunt: aut fortitudo, aut velocitas* (Barney et al. 2006, 253, XII.II.26; Lindsay ed. 1971, XII.II.26). In literary evidence and chronicles, horses are regularly praised concerning their swiftness (Walker-Meikle 2017; Ropa 2019). This applies, in particular, to war horses and especially the Turkish horses that could, in German literature, be called simply *schnelle Türken* (swift Turks) as, for instance in the heroic Upper German epic *Ortnit* (ca 1230): *Einen snellen türcen der Lamparte twanc....* (Amelung and Jänicke eds 1968, 41, stanza 310; see Jaritz 2002, 42).

The discussion about the velocity or slowness of animals could, however, become much more detailed, containing comparisons to other animals and also be connected with human qualities and positive or negative human roles. Most such information can be found in the bestiaries, particularly popular in France and England in the twelfth century. "If one brings together medieval literary history and animal studies, one major point of juncture between the two is the genre of bestiary" (Kay 2017, 20). Based on the early Christian *Physiologus*, a text written or compiled in Greek in the second or third century CE, bestiaries not only offer descriptions and visual representations of animals but also often use them, in particular, for Christian moralization. They have been seen as didactic sermon source material for priests (Dykema 2011) or as elementary teaching material aimed at lay brothers or schoolboys (Kay 2017, 13–15). Some examples of swift and slow animals in them will be offered from the well-known *Aberdeen Bestiary* from about 1200.

The *leucrota*, a composite animal from India, is sometimes said to have been as big as an ass, with the hindquarters of a stag, the chest and legs of a lion, the head of a horse, and exceeding in velocity all existing wild animals: *De leucrota. In India nascitur bestia nomine leucrota, que velocitate precedit feras universas. ...* (*Aberdeen Bestiary*, fol. 15v; Clark 2006, 139; Barber 1999, 61). This was taken over from Pliny's *Historia naturalis*: ... *leucrocotam pernicissimam feram* ... (Rackham 1983, 54, book viii, 30; see also Boese ed. 1973, 146).

The comparison of the efficacy of different animal qualities also appears with regard to the lion. There, the discussion concerns the speed of other animals and how they cannot outrun the lion. Thomas of Cantimpré refers in this respect to Saint Ambrose, asking:

> Which beast dares to rouse the lion, whose voice, by its nature, inspires such terror that many living things which could evade its attack by their speed, grow faint at the sound of its roar as if dazed and overcome by force?
>
> *Que autem ei* [*leoni*] *se cire fere audeat, cuius voci tantus naturaliter inest terror ut multa animantium que per celeritatem possunt evadere eius impetum, rugitus eius sonitum velud quadam vi attonita atque victa deficiant?* (*Aberdeen Bestiary*, fol. 8r; Boese ed. 1973, 140)

Fast horses become part of the description about the sirens, not the sea creatures which typically take the form of a half-human with a fish tail, but snakes living in Arabia, about which the Aberdeen Bestiary says that they are white and cover the ground faster than horses and are also said to fly: *De sirenis. In Arabia autem serpentes albi sunt cum alis, que sirene vocantur, que plus currunt ab equis, sed etiam et volare dicuntur* (*Aberdeen Bestiary*, fol. 69v). This time, the description is based on Isidore of Seville: *In Arabia autem sunt serpentes cum alis, quae sirenae vocantur, quae plus currunt ab equis, sed etiam et volare dicuntur* (Barney et al. 2006, 258, XII.iv.29; Lindsay ed. 1971, XII.iv.29). Another comparison of velocity and slowness refers to the caterpillar (*eruca*) and the locust: The caterpillar folds itself up and does not fly about like the locust, which hurries from place to place, in all directions, leaving things half-eaten, but stays amid the fruits that are destined to be destroyed and, munching slowly, consumes everything: *Eruca. ... Implicat se idem nec advolat, ut locusta huc illucque discurrens, semipasta dimittit, sed permanet perituris frugibus et tardo lapsu et pigris morsibus universa consumit* (*Aberdeen Bestiary*, fol. 72r).

A more detailed comparative discussion can be traced with regard to the ostrich in contrast to gyrfalcons and hawks. Again following the Aberdeen Bestiary:

> The wings of the ostrich resemble those of the gyrfalcon [the largest of the falcon species] and the hawk. Who does not know how the speed of the gyrfalcon and hawk in flight exceeds that of other birds? The ostrich certainly has wings like theirs but not their speed of flight.... In considering the ostrich, we should look more carefully at the hawk and the gyrfalcon. Their bodies are small but their feathers are more densely packed; as a result, they fly at great speed, because they have little to weigh them down, much to uplift them. In contrast, the ostrich has few feathers and is weighed down with a huge body, so that even if it were to try to fly, its sparse feathers would not support the mass of such a large body in the air.
>
> *Penna strucionis similis est pennis herodii et accipitris. Quis herodium et accipitrem nesciat, aves reliquas quanto volatus sui velocitate transcendat? Strucio vere penne eorum similitudinem habet, sed volatus eorum celeritatem non habet. ... Habemus quod in considerationem strucionis huius de accipitre et herodio attentius perpendamus. Accipitris quippe et herodii parva sunt corpora sed pennis densioribus fulta, et iccirco cum celeritate transvolant, quia eis parum inest quod aggravat, multum quod levat. At contra strucio raris pennis induitur, et immani corpore gravatur, ut et si volare appetat ipsa pennarum paucitas molem tanti corporis in aere non suspendat.* (*Aberdeen Bestiary*, fol. 41v)

This example of ostrich versus gyrfalcon and hawk continues and also leads to one additional, already mentioned aspect of the perception of (fast and slow) animals in bestiaries which often plays a particular role in their descriptions, that is, their moralistic and didactic comparison with humans and their qualities:

> Truly, the ostrich has not the capacity to be lifted from the ground and gives only the impression of spreading its wings as if to fly; however, it never supports itself above the earth in flight. It is exactly the same with all those hypocrites who pretend to live a life of piety, giving the impression of holiness without the reality of holy behaviour. They certainly have wings, as far as appearance goes, but in terms of action, they creep along the ground, because they spread their wings only to give an illusion of holiness, but they cannot possibly raise themselves from earth, weighed down as they are by the weight of worldly preoccupations. ... The gyrfalcon and the hawk, however, represent the elect well. Who, in this life, are not without the contami-

nation of sin, no matter how trivial. But when the very small amount of sin that is within them causes them to sink, the large amount of virtue amassed from their good works is at hand to lift them back up to the heights. ... The large number of the hypocrite's bad deeds, compared to his too few good deeds, weigh him down.

A terra quippe elevari non valet, et alas quasi ad volatum specie tenus erigit, sed tamen nunquam se a terra volando suspendit. Ita sunt nimirum omnes ypocrite qui dum bonorum vitam simulant, imitationem sancte visionis habent, sed veritatem sancte actionis non habent. Habent quippe volandi pennas per speciem, sed in terram repunt per actionem, quia alas per figuram sanctitatis extendunt, sed curarum secularium pondere pregravati, nullatenus a terra sublevantur .... Bene ergo in herodio et acciptre electorum persona signatur, qui quamdiu in hac vita sunt sine quantulocumque culpe contagio esse non possunt. Sed cum eis parum quid inest quod deprimit, multa virtus bone actionis suppetit, que illos in superna sustollit ... parum bonum ypocrite multitudo prave actionis gravat. (*Aberdeen Bestiary*, fol. 41v–42r)

These and all other moralistic and didactic animal–human comparisons, using animals as models, are to be seen as based on a general statement with regard to birds that certainly can be extended to all animals, saying, "See how birds who lack the capacity of rational thought instruct through examples of evil conduct, men who are experienced and intelligent" (*Ecce quomodo volucres que ratione carent peritos homines et ratione intentos per exempla perverse operationis docent*; *Aberdeen Bestiary*, fol. 46v).

As already mentioned, Isidore of Seville stated that the vulture got its name from *volatu tardus*, slow in flight. Similar to the ostrich example, the Aberdeen Bestiary adds that the vulture leaves the ground slowly when it takes flight, as the sinner hardly ever or never abandons his earthly desires: *Tarde enim cum volare ceperit a terra recedit, quia peccator aut vix aut nunquam terrena desideria derelinquit* (*Aberdeen Bestiary*, fol.45v). In other bestiaries, the vulture also represents laziness, because it flies slowly (Clark 2006, 20).

Such a connection of tardiness of animals and laziness can also be found with regard to the ass in the literary and visual representations and personifications of sloth as one of the Seven Deadly Sins (Wenzel 2012). The laziness of the ass in connection with human behavior may already be found in texts from the sixth century CE in Boethius' (480/485–524/526) *Consolatio Philosophiae*: [*Homo*] *Segnis ac stupidus torpet? Asinum vivit* (Weinberger ed. 1934, 87; Goins and Wyman eds 2012, 74; Sedgefield 1900, 132). Isidore of Seville states that it is a slow animal and balks for no reason: *Asinus ... animal quippe tardum et nulla irratione renitens ...* (Barney et al. 2006, 249, XII.i.38; Lindsay ed. 1971, XII.I.38–39). Particularly during the Late Middle Ages, this tardiness was adopted as the textual and visual personification of sloth. Although certainly not exclusively, the ass became the standard animal symbol for *acedia* (Wenzel 2012). In the Flemish mystic Willem Jordaens' (ca 1321–1372) debate poem *Conflictus virtutum et viciorum*, the Virtues use animal names as swear words against the Vices, that is, "ass" against sloth: *rudis asina* (Önnerfors ed. 1986, 102, n. 189). The didactic visual images of the deadly sins also regularly show an ass as part of the personified sin of sloth (Blöcker 1993; O'Reilly 1988). For instance, an end-fourteenth-century wall painting in the parish church of Levoča in Slovakia represents sloth's personification as a sleeping couple, with their heads on cushions, ride on an ass into the Mouth of Hell (Fig. 4.1).

The swiftness or slowness of animals could also be seen as dependent on the occasion and, thus, become a model for a situation contingent on the

Fig. 4.1 The ass as part of the personification of the capital sin of sloth. Wall painting, end of the fourteenth century, parish church St. James, Levoča (Slovakia). (Photo: Institut für Realienkunde, University of Salzburg (Austria))

behavior of humans. Such an example of occasion-dependent speed is offered by the free-range goat:

> About the goat: ... When grazing, it moves from high to even higher pastures. ... When wounded, it hurries to find the herb dittany [*Origanum dictamnus* (dittany of Crete)] and, by touching it, is healed. In the same way, good preachers graze on the law of the Lord and take delight in good works as in good pastures, rising from one virtue to another. ... When wounded by sin, they hurry back to Christ by confessing and are quickly healed. For this reason, Christ is rightly said to be like dittany. For as dittany drives out iron from a wound and heals it, so Christ through confession casts out the devil and pardons sin.
>
> *De capra: ... habet naturas, quod pascendo de altis ad altiora tendit. ... Vulnerata ad ditannum currit, qua tacta sanatur, Sic boni predicatores, pascentes in lege domini, et in bonis operibus quasi in bono pastu delectantes, de virtute in virtutem conscendunt. ... A peccato vulnerati, ad Christum confitentes recurrunt, et cito sanantur. Ideoque Christus bene ditannus dicitur. Sicut enim ditannus ferrum a vulnere depelit, et vulnus sanat, ita Christus per confessionem diabolum eicit, et peccatum ignoscit.* (*Aberdeen Bestiary*, fol. 14v)

The *Concordantiae caritatis* is one of the most important typological image-text-cycles of the Middle Ages, again to be seen as didactic sermon source material for priests. Its author was Ulrich von Lilienfeld, monk of the Austrian Cistercian house of Lilienfeld, where he served as abbot from 1345 to 1351, the year he resigned, perhaps to concentrate more on his literary activities. He died sometime before 1358. The *Concordantiae caritatis* was written in the 1350s and explains with images and commenting texts, the gospels of the church year and legends of saints with typological scenes from the Old Testament and examples out of nature (Douteil 2010). The latter refer mainly to animals, their characteristics, and activities. Some of them also relate to the swiftness of animals. In the biblical story of Jesus Walking on the Sea, for instance, the author sees the sea as an image of the world and its disquietude. Jesus walked over the sea, as his activities were beyond this world. This could be compared to the fish *ludolacra* (a flying fish; Fig. 4.2). Based on Thomas of Cantimpré's *Liber de natura rerum* (Boese ed. 1973, 243), Ulrich von Lilienfeld states:

> Aristotle says that the ludolacra is an animal living in the sea; it has four fins, two on the head and two on the back, and with them it crosses the sea very quickly. The same goes for you: If you want to cross the stormy sea of this

Fig. 4.2 *Concordantiae caritatis*, Library of the Cistercian house of Lilienfeld (Lower Austria), cod. 151, fol. 32v (detail), the *ludolacra*: *Ad loca longa salis, ludolacra, quatuor alis* (Ludolacra, with four fins, you leap to distant places). (Photo: Institut für Realienkunde, University of Salzburg (Austria))

world safely, you must have two fins on your head: one of them is pious prayer, the other constant contemplation of heavenly things, and two on the back, that is, mortification of carnal desire and almsgiving. With these fins and with the help of true cognition of the Lord you will very safely and quickly overfly the deceptive temptations of this worldly sea towards the longed for place, that is, the presence of God (transl. author):

Aristotiles dicit, quod ludolachra animal marinum IIIIor habet alas, duas in facie et duas in dorso, et hiis velocissime mare transit. Ita tu uolens secure huius mundi tempestuosum transire mare necessarie est, ut duas habeas alas in facie, vnam deuotam oracionem, aliam iugem de celestibus contemplacionem, et duas in dorso, scilicet carnis mortificacionem et elemosinarum largicionem, et hiis alis in vera Domini adiuuante cognicione huius maris mundialis lubrica temptamenta securissimus at velocissimus usque ad locum optatum, id est diuinam presenciam transuolabis. (Douteil 2010, vol. 1, 68)

A more familiar swift animal is the squirrel. Ulrich von Lilienfeld puts it into typological context with Saint Hippolytus (Fig. 4.3):

> The squirrel is a small and swift animal that collects nut kernels in the summer to eat them in winter, and it flies with its tail. The blessed Hippolytus was a squirrel, small in his humility and swift in his service to God. ... And from one place to the other, that is, from the present epoch to heavenly joy, he flew away with his tail most joyfully, that is, with the steadfastness of his passion. Thus, we also should swiftly collect sweet merits in this life to sense the most pleasant rewards in the future of eternal blessedness. (transl. author)
>
> *asperiolus animal paruum et agile nucleos in estate congregat, quos in hyeme comedat, et cauda uolat. Animal asperiolus fuit beatus Ypolitus, paruum humilitate, agile in diuina seruitute, ... et de loco ad locum, id est de presenti seculo ad celeste gaudium cauda, id est cum passionis perseuerancia*

Fig. 4.3 *Concordantiae caritatis*, Library of the Cistercian house of Lilienfeld (Lower Austria), cod. 151, fol. 203v (detail), the squirrel: *Cauda volat solus fit et agilis asperiolus* (The swift squirrel flies just with its tail). (Photo: Institut für Realienkunde, University of Salzburg (Austria))

> *letissimus auolauit. Sic et nos in hac uita agiles nobis dulcia merita conportemus, vt in futuro eterne beatitudinis suauissima premia senciamus.* (Douteil 2010, vol. 1, 426)

The swiftness of the eel occurs in another context (Fig. 4.4). Based on Saint Gregory, Ulrich von Lilienfeld states in connection with the Parable of the Friend at Night (Luke 11; 5–8):

> Saint Gregory writes: The more the eel is squeezed, the quicker it wrests itself free; and also the other way round. It happens like that with this world; it is the worst eel. If somebody wants to squeeze it intensely, that is, use it at will and followed the comforts of its effeminacy, it quickly slips from him either through the aggressiveness of death or the hot snare of temptations and cannot, as he wanted, be retained at will. (transl. author)

Fig. 4.4 *Concordantiae caritatis*, Library of the Cistercian house of Lilienfeld (Lower Austria), cod. 151, fol. 114v (detail), the eel: *Liberat angwillam, si stricta manus tenet illam* (When holding it strongly, the hand frees the eel). (Photo: Institut für Realienkunde, University of Salzburg (Austria))

> *Beatus Gregorius scribit, quod anguilla quanto plus premitur, tanto cicius elabitur, et conuerso. Sic hic mundus, qui est anguilla pessima, quam si quis fortiter conprimere uoluerit, id est eo ad libitum uti et eius mollicie dulcedinem secutus fuerit, ab illo aut per mortis uiolenciam aut per temptamentorum callidam decipulam cito prolabitur nec, prout uellet, ad placitum retinetur.* (Douteil 2010, vol. 1, 232)

The discussion of slowness and speed of representatives of the faunal world could, in these contexts, play a significant role that not only helped the medieval audience to understand parts of nature but, in particular, one's own self. The latter aspect even led to a statement in one of the mystical writings of Meister Eckhart (ca 1260–1327/1328), *Von der Abgeschiedenheit* (On detachment):

> Now take note, all who have good sense! The swiftest animal that bears you to His perfection is suffering, for none will enjoy greater eternal bliss than those who stand with Christ in the greatest bitterness. Nothing is more gall-bitter than suffering, nothing more honey-sweet than having suffered.
>
> *Nû merket alle vernünftigen menschen! Daz snelleste tier, daz iuch treget ze dirre volkomenheit, daz ist lîden, wan ez geniuzet nieman mê êwiger süezicheit, dan die mit Kristô stânt in der grœsten bitterkeit. Ez enist niht gelligers dan lîden und enist nicht honicsamers dan gelitten hân.* (Walshe ed. 2009, 574; Largier ed. 1993, 458. In the English translation of Walshe ed. 2009, 574, the original *tier* (= animal) was rendered as "steed.")

All these examples show that the medieval interest in and perception of animals could lead in a variety of different directions, from becoming aware of known and unknown phenomena in the natural world to recognizing them as positive or negative models and comparative patterns for one's own life. In many respects, comments on the swiftness and slowness of animals offered such information as well as role models.

References

Primary Sources

Aberdeen Bestiary. https://www.abdn.ac.uk/bestiary/. accessed August 18, 2020.

Amelung, Arthur, and Oskar Jänicke, eds. 1968. *Ortnit und die Wolfdietriche nach Müllenhoffs Vorarbeiten*, vol. 1. Berlin: Weidmannsche Buchhandlung, 1871; reproduction: Dublin and Zürich: Weidmann.

Barber, Richard, transl. 1999. *Bestiary: Being an English Version of the Bodleian Library, Oxford, MS Bodley 764.* Woodbridge: Boydell.

Barney, Stephen A., Wendy J. Lewis, Jennifer A. Beach, and Oliver Berghof transl. 2006. *The Etymologies of Isidore of Seville.* Cambridge: Cambridge University Press.

Boese, Helmut, ed. 1973. *Thomas Cantimpratensis liber de natura rerum*, vol. 1: *text.* Berlin and New York: Walter de Gruyter.

Douteil, Herbert. 2010. *Die* Concordantiae caritatis *des Ulrich von Lilienfeld: Edition des Codex Campililiensis 151 (um 1355)*, 2 vols. Münster: Aschendorff.

Goins, Scott, and Barbara H. Wyman, eds/transl. 2012. *Boethius, The Consolation of Philosophy.* San Francisco: Ignatius Press.

Largier, Niklas, ed. 1993. *Meister Eckhart, Werke* II. Berlin: Deutscher Klassiker Verlag.

Lindsay, Wallace Martin, ed. 1971. *Isidori Hispalensis episcopi etymologiarum sive originum libri XX*, vol. 2: *Libros XI–XX.* Oxford: Clarendon Press (1911; repr).

Önnerfors, Alf, ed. 1986. *Willem Jordaens Conflictus virtutum et viciorum.* Opladen: Westdeutscher Verlag.

Rackham, Harris, transl. 1983. *Pliny Natural History*, vol. 3, 2nd ed. London–Cambridge MA: William Heinemann–Harvard University Press.

Sedgefield, Walter John, transl. 1900. *Boethius, King Alfred's Version of the Consolations of Boethius.* Oxford: Clarendon Press.

Walshe, Maurice O'C., ed./transl. 2009. *The Complete Mystical Works of Meister Eckhart.* New York: Crossroad.

Weinberger, Guillelmus, ed. 1934. *Anicii Manlii Severini Boethii Philosophiae Consolationis libri quinque.* Vienna–Leipzig: Hölder-Pichler-Tempsky–Akademische Verlagsgesellschaft.

Secondary Sources

Blöcker, Susanne. 1993. *Studien zur Ikonographie der Sieben Todsünden in der niederländischen und deutschen Malerei und Graphik: von 1450–1560.* Münster: LIT.

Clark, Willene B. 2006. *A Medieval Book of Beasts: The Second-family Bestiary: Commentary, Art, Text and Translation.* Woodbridge: Boydell.

Dykema, Bobbi. "Preaching the Book of Creation: Memory and Moralization in Medieval Bestiaries." *Peregrinations*: Journal of Medieval Art and Architecture 3,2 (2011). 96–121.

Jaritz, Gerhard. "Fear and Fascination: Late Medieval German Perceptions of the Turks Revisited." *Medium Aevum Quotidianum* 46 (2002). 40–47.

Kay, Sarah. 2017. *Animal Skins and the Reading Self in Medieval Latin and French Bestiaries.* Chicago and London: The University of Chicago Press.

Lenoir, Frédéric. 2017. *Lettre ouverte aux animaux (et à ceux qui les aiment)*. Paris: Arthème Fayard.

O'Reilly, Jennifer. 1988. *Studies in the Iconography of the Virtues and Vices in the Middle Ages*. New York: Garland.

Ropa, Anastasija. 2019. *Practical Horsemanship in Medieval Arthurian Romance*. Budapest: Trivent.

Walker-Meikle, Kathleen. 2017. *The Horse Book: Horses of Historical Distinction*. London: Bloomsbury.

Wenzel, Siegfried. 2012. *The Sin of Sloth: Acedia in Medieval Thought and Literature*. Chapel Hill: University of North Carolina Press.

CHAPTER 5

Animals Between Authors and the Natural World in Giovanni da San Gimignano's *Liber de exemplis et similitudinibus rerum*

Beatrice Amelotti

The aim of this chapter is to briefly present the figure of the Dominican preacher Giovanni da San Gimignano and his moralizing encyclopedic work written around 1300, the *Liber de exemplis et similitudinibus rerum*, focusing on a particular point: how and why did this preacher bring and apply zoological *exempla* and *similitudines* into his sermons. How did he use animal attributes to define and explain the moral qualities needed in a preacher operating at the beginning of the fourteenth century in central Italy. Since the fundamental key to contemporary validation of encyclopedias of the Middle Ages lies in the references that are always made to past authorities (*auctoritates*) through quotations from the works of other authors, the original sources have been listed wherever possible.

Giovanni da San Gimignano was born in San Gimignano between 1260 and 1270. He lived until at least 1333, therefore making him a contemporary of Dante Alighieri. He entered the Dominican order in Siena and actively preached throughout central Italy. It appears likely that he also

B. Amelotti (✉)
Dipartimento di Studi Umanistici, Università degli Studi di Pavia, Pavia, Italy

L. Bartosiewicz, A. M. Choyke (eds.), *Medieval Animals on the Move*, https://doi.org/10.1007/978-3-030-63888-7_5

studied abroad for some period of time, probably at the *studium generale* in Barcelona. There are various mentions of the contacts he is known to have had with people from the Arab world (Dondaine 1939, 131), a distinct possibility in the rapidly expanding Principality of Catalonia. After his training in Barcelona came to an end, Giovanni was appointed prior of a convent in Siena and, together with two other brethren, he drew up the statutes of the *Confraternita della Beata Vergine e San Domenico* (Confraternity of the Blessed Virgin of Saint Dominic). Last but not least, between 1318 and 1329, he was among the founders of the first Dominican convent in the small walled medieval hill town of San Gimignano in the province of Siena in Tuscany.

Aside from his most important work *Liber de exemplis et similitudinibus rerum* (*LdE* henceforth), Giovanni mostly wrote collections of sermons, four of which (*Sermones de tempore*, *Sermones de sanctis*, *Sermones funebres*, and a *Quadragesimale*) can be attributed to him with certainty. In addition to this genre, he is also author of a hagiographical work, the *Legenda Sanctae Finae* (The legend of Holy Fina; for further details on the life and works of Giovanni da San Gimignano, see also: Quétif, Échard 1719, I, 528 s.; II, 819; Stegmüller 1951, 420–422; Raffin 1974, coll. 721 s.; Kaeppeli 1975, II 539–543, 1993, IV 170; Oldoni 1994, 212–228; Kaeppeli 1993, IV, 170; Vecchio 2001, 206–209).

Unlike the collections of sermons, which are consistently attested in Italy, the 72 manuscripts comprising his encyclopedia are mostly held in Central and East European libraries, that is, in East Germany, Poland, the Czech Republic, and Hungary (Van den Abeele 2002, 300–304). This is probably a consequence of Dominican expansion into those regions and their need for proper preaching instructions and related materials. This most interesting phenomenon is also typical of other encyclopedias such as the work of Bartholomaeus Anglicus (Bartholomew the Englishman), an early thirteenth century Parisian Franciscan scholar who also discussed the attributes of animals in the 1240 compendium *De proprietatibus rerum* (On the Properties of Things), an early forerunner of the encyclopedia format and a widely cited book in the Middle Ages.

Before returning to the topic of this chapter, a brief introduction to medieval encyclopedias and their purpose will be useful. What precisely was a medieval moralizing encyclopedia and what were the purposes it served? These encyclopedias were basically works drawing their materials from common encyclopedias, texts intended for distinctly educational purposes and providing descriptions of different aspects of the natural world; some very well-known examples include the previously mentioned

De proprietatibus rerum by Bartholomaeus Anglicus and the thirteenth century *Speculum Maius* (Great Mirror) by the French Dominican, Vincent de Beauvais. The aim of a moralizing encyclopedia was not only to treat and describe the variety of the Creation as well as various elements of the natural world, humans included, but also to teach and indicate proper behavior for good Christians. Therefore, the various features and attributes known to be possessed by plants, stones, animals, and so on were used to show, through similes and analogies, models to be followed or, if these characteristics are negative, examples of what behavior should be avoided.

Above all, a boundless series of healthy moral lessons were offered by the animal world. Animals from everyday life and those never seen, such as exotic animals from far-off places mentioned in various sources.

In order to understand the significance of this genre, it is important to clearly identify its audience. From the thirteenth century, a moralizing encyclopedia, exactly like an ordinary encyclopedia, was generally written by and addressed to preachers, mostly Dominicans and Franciscans, who looked for inspiration in those works during the writing of their own sermons. The material drawn from the encyclopedias was used, in fact, to illustrate the Holy Bible and to better convey its sometimes abstract concepts to lay readers.

The primary goal of Giovanni's *LdE* was to give his brethren a useful preaching aid, a source from which they could extract material to create successful sermons, that is, sermons with a great public appeal. This work consists of ten books, all rich in *similitudines*, and each treating one particular sphere in God's Creation. It starts with the Heavens, passing through minerals and metals, and ends with the products of human endeavor. As mentioned, Books IV and V are devoted to animals, each divided into two large groups. In the first of these two books, the protagonists are fish and birds, and in the second, a variety of quadrupeds not included in book IV. The examples of animals are introduced by two prologues that illustrate the multiple relationships between humans and the animal world.

The animals are always used as images, similes for something else, following the concept of Christian Platonism, according to which, the world of the senses reflects a superior and divine reality. Consequently, the study of living animals, along with all the natural sciences, was subordinated to theology. Moving animals from the natural world and using attributed characteristics functioned to aid a deep comprehension of the Bible, as clearly stated by St. Augustine in his *Epistulae* LV, 7, 12 (Zambon 2001, 31) and by St. Bonaventura in his *Collationes in Hexaëmeron*, 150, visio III, collatio I.

There are nine chapters in which preachers are mentioned. Of these nine chapters, I have selected three passages from Book V, chapter 98. In this part of the work, *similitudines* between the preacher and three well-known animals transiting from the natural world surrounding people, the dog, the beaver, and the bear, are illustrated as similes transferred in a moralizing form to the written page.

As anticipated, when considering the genre of medieval encyclopedias, apart from examining the content of the passage, it is always worth looking at the potential sources. Medieval authors always quoted and referred to various authorities (*auctoritates*) in their writing, that is, these authors were not really original but attempted to convey their own thoughts through the words of their best-known predecessors. Nevertheless, it is often not easy to retrace the sources of the various quotations since the original author's name is rarely mentioned. Even when it is, the reference is often made to an indirect source, namely it is quoted through another author's work. For example, Giovanni da San Gimignano frequently refers to Aristotle, but his knowledge of the Aristotelian doctrine is indirect, actually coming from the works of Thomas Aquinas. Thus, these animals not only move from their natural state to the page, but the notions about their attributed characteristics also move as they were borrowed from medieval work to medieval work.

Therefore, using the *Brepolis Cross Database Searchtool* (CDS, Brepols Publishers. ‹https://about.brepolis.net/cross-database-searchtool/, accessed 04/07/2018), the *Corpus Corporum* by the University of Zürich (*Corpus Corporum. Repositorium operum Latinorum apud universitatem Turicensem*. Database under development. ‹http://www.mlat.uzh.ch/MLS/index.php?lang=0›, accessed 04/07/2018), *Sermones.net* (*Sermones.net. Édition électronique d'un corpus de sermons latins médiévaux*. Directed by Nicole Bériou. ‹http://sermones.net/›, accessed 04/07/2018) and the *Thesaurus Exemplorum Medii Aevi* (ThEMA, directed by Jacques Berlioz, Marie Anne Polo de Beaulieu, Pascal Collomb. ‹http://gahom.ehess.fr/index.php?434›, accessed 04/07/2018), when an *auctoritas* was directly mentioned in the text, I verified the provenience of the quotation, sometimes even discovering intermediary sources. However, there were cases where the preacher did not indicate any *auctoritas*, and it proved impossible to identify the possible source. In these cases, it may well be possible that the relevant text was either Giovanni da San Gimignano's original or that the works he consulted have not yet been included in the aforementioned databases and are therefore not currently traceable.

The first simile presented in detail here connects preachers and dogs: *predicator assimilatur cani.* ("A preacher is like a dog.") The association of the Dominicans with dogs is well known, since it plays an important role in their iconography. The Dominicans, according to the popular interpretation of their name were *domini canes* (the Lord's dogs), traditionally represented as black and white dogs, reflecting the black and white colors of the order's habit. A famous example is the fresco *Esaltazione dell'ordine domenicano* (Exaltation of the Dominican order, 1365–1368) by Andrea di Bonaiuto, which can be seen in the Cappellone degli Spagnoli (the Spanish Chapel) of Santa Maria Novella in Florence (Fig. 5.1). Why were the preachers identified with dogs? What were thought to be the common features between the *fratres* and this ordinary domestic animal?

First, a dog loves his owner unconditionally. In the same way—Giovanni says—the preacher is supposed to love God, and this unconditional love is the prerequisite for preaching the *verba Dei*, the words of God. All the quotations from the *LdE* which follow are cited from the 1609 Anvers edition:

> *Predicator assimilatur cani. Primo ex eo, quia canis dominum suum diligit:* [...] *predicator qui dominum non diligit non est dignus annunciare verba Dei.* (*LdE* V, 98, f. 273v)
>
> The preacher is similar to the dog. Firstly, because the dog loves his owner: [...] a preacher who does not love God is not worthy of announcing God's message.

Secondly, a dog typically protects its home, and the preacher should do the same thing with the Church, the dwelling of God:

> *Canis Domini sui custodit domum, et defendit. Hec domus Domini est Ecclesia, quam Predicator Catholicus ab hereticis defendit.* (*LdE* V, 98, f. 274r)
>
> God's dog guards and defends his house. God's house is the Church, which the Catholic preacher protects from heretics.

Analyzing the structure of this simile, it can first be seen in terms of the comparison itself. Giovanni then describes the characteristics of the

Fig. 5.1 Dominicans and dogs with their black and white habits depicted by Andrea di Bonaiuto, Cappellone degli Spagnoli, Santa Maria Novella, Florence. Detail. (Wikimedia commons. Image in the public domain). https://commons.wikimedia.org/wiki/File:Andrea_Bonaiuti-The_Church_as_the_Path_to_Salvation_(detail_).jpg

animal, which, in this case, reflect medieval stereotypes of canine behavior. According to the common use of *similitudines*, these characteristics can also be physical ones. Finally, as Giovanni builds the analogy between the dog and the preacher, he links the typical behavior attributed to dogs to the duty of the human.

In this particular case, the precise match between the text of the *LdE* and other works is still unclear. This does not mean, of course, that there are no other occurrences of Dominican friars being compared to dogs. As a matter of fact, mentioning emphatically positive dog behavior was a very commonly used analogy in the Middle Ages, as shown by this example from the following passage by Pope Gregory I (c. AD 540–12 March 604), also known as *Gregorius Magnus* (Saint Gregory the Great) in his *Homiliae in euangelia* (Homilies on Gospels) II:

> *Nonnumquam solent in sacro eloquio per canes praedicatores intelligi. Canum etenim lingua vulnus dum lingit curat, quia et doctores sancti dum in confessione peccati nostri nos instruunt, quasi vulnus mentis per linguam tangunt,*

> *et quia nos loquendo a peccatis eripiunt, quasi tangendo vulnera ad salutem reducunt.* (Gregorius Magnus, *homilia* 40, 35)
>
> In the Holy Scripture preachers are sometimes referred to as dogs. In fact, the dog's tongue cures wounds by licking them; in the same way, during the confession of our sins, holy doctors elevate us, it is like they touch the spiritual wound with their tongues, and since they tear us away from sin, they draw us to salvation by touching the wounds, as it were.

It is quite easy to find other similes where the protagonists are preachers and dogs, but none of the passages I found can be considered the source for Giovanni da San Gimignano since the discrepancies between those texts and the chapters of the *LdE* are too great. After having reviewed many cases of *similitudines*, the following text represents a passage containing an *exemplum*:

> *Multotiens canis corpus domini sui etiam mortuum non derelinquit, sicut legimus de cane Titii Sabini, qui ipsum Dominum suum captum Rome etiam in carcere non dereliquit: quo insuper mortuo, canis mestos edens ululatus, nec mortuum deseruit; qui, cum quidam ei panem proiecisset, ad os defuncti deportavit* [...], *nec ipse manducavit. Cadavere autem in Tyberim proiecto, canis, ut corpus sustentaret, in Tyberim se proiecit* [...]. *Similiter predicatores corpus Christi mortuum deserere non debent, quia de morte et passione Christi debent iugiter memoriam habere, et de ea predicare frequenter.* (*LdE* V, 98, f. 274r)
>
> Many times, a dog does not leave its master even after his death, as we read about Titius Sabinus' dog, who did not abandon its owner after he was imprisoned in Rome, and even after his death it kept on howling sadly next to its master's corpse. When someone threw it some bread, it brought it to his owner's mouth without eating it itself [...]. When the corpse was thrown in the river Tiber, the dog swam to it in order to keep it afloat [...]. In the same way, preachers should not leave Christ's corpse and forget about Christ's passion and death; instead, they are supposed to preach frequently about them.

This *exemplum* illustrates the fact that dogs never abandon their owners, even after the master's death. The story is about the dog of Titius Sabinus, whose master, according to Plinius Maior (Pliny the Elder; *Naturalis Historia* [Natural History] book 8, 145), was imprisoned because of his opposition to Emperor Nero. After the death of Titius Sabinus, his faithful dog kept watch over him, even bringing him food. However, as also seen before, a search through key-words in the *CDS*,

Corpus corporum, and *Thema* yielded no evidence of a possible source used by Giovanni da San Gimignano.

As can be seen from the text, the *exemplum* is a short story, presented as true, whose aim was both to teach and entertain (*docere et delectare*) at the same time (Von Moos 1992, 67). The behavior of the dog becomes, once more, a model for the preacher.

After the short *exemplum*, the *similitudo* with the beaver is worth considering. In contrast to the previously discussed commonly known dogs, a detailed description is dedicated to these animals which by this time had become less common in Europe. Unlike the dog, Giovanni may never actually have seen a beaver. The beaver thus moves from the biological animal into the world of imagination. Giovanni's beaver is said to be as big as a small dog and to have canine back paws, while its front paws can be likened to the feet of a duck (which is just the opposite of anatomical reality: Fig. 5.2). In this way, the beaver can live and move both in the water and on land. The information about the beaver is attributed to Isidorus Hispalensis (Isidore of Seville, c. AD 560–4 April 636):

> *Item Predicator debet similis esse castori, qui (sicut dicit Isidorus) cum quadrupedibus vivit in terra, et cum natatilibus stat in aquis.* (*LdE* V, 98, f. 274v)
>
> Then the preacher must be similar to a beaver, which (as Isidorus says) lives both on the earth with quadrupeds and in the water with aquatic animals.

Turning to the analogy, in this case, land is seen as the active life (*vita activa*) and water is seen as the contemplative life (*vita contemplativa*): the good preacher should illustrate the *Sacra Scriptura* (Sacred scriptures) to both active and contemplative characters.

It should be noted that in *LdE* V, 17, the *similitudo* with the beaver is used to describe the virtues of people who live in chastity and abstinence (*castitate et continentia*), people who are, therefore, able to keep sin away from themselves through self-control. This double use of the same animal to convey two different concepts should not surprise the reader. In fact, it is very common to read different similes derived from the same object, whether an animal, a plant, a precious stone, or anything else. Sometimes the same element can represent two opposite features, one positive and one negative (Ledda 2003, 58). Once again, the *auctoritas* of Augustinus Hipponensis (St. Augustine of Hippo) can be useful in understanding this dualism more deeply. In fact, in his *De doctrina christiana* (On the Christian Doctrine III, 25, 36), Augustinus clearly states that each

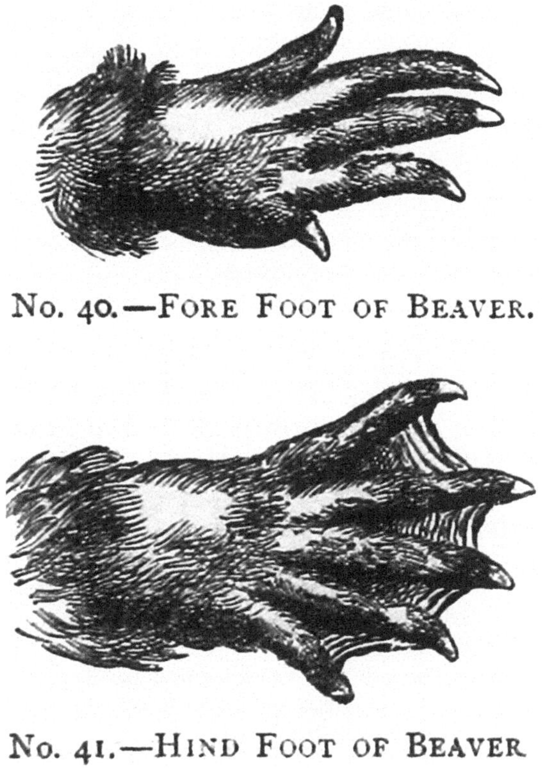

Fig. 5.2 Nineteenth century scientific drawing of beaver feet (Boutell 1869, p. 23 No. 40–41. Wikimedia commons. Image in the public domain). https://commons.wikimedia.org/wiki/File:Picture_Natural_History_-_No_40_41_42_-_Beaver_feet_and_tail.png

element in the Bible can have different or even opposite meanings. The lion, which represents both Christ and the Devil (Zambon 2001, 36–37), is one such example. As the Bible represented the universal model to follow at the time, it is clear that the same structure could be commonly employed.

Referring to the two similes inspired by the beaver, it may be worth stressing that despite this being the same animal, the traits attributed to it are different, as was the source, which in *LdE* V, 17 can be probably identified in as in *De proprietatibus rerum* (XVIII, 28) by Bartholomaeus Anglicus.

The last *similitude* discussed here is one where the preacher and the bear are compared:

> *Item Predicator etiam assimilatur urso. Nam quia ursus sic interpretatus et dictus est; quod ore suo formet fetus (ut dicit Isidorus). Avicenna etiam dicit quandam carnem nasci quam mater lambendo in membra componit. Sic etiam Predicator doctrina oris sui, format Deo spirituales filios.* (*LdE* V, 98, f. 274v)
>
> Thus, the preacher is also similar to a bear. The bear is called this way because with its mouth—*ore suo*—it shapes the fetus (as Isidorus says). Avicenna also says that the mother bear licks the cub and gives it a shape after its birth. In the same way, the preacher, with the doctrine of his mouth, forms spiritual sons for God.

The animal's name was supposed to derive from the fact that, according to a widely held popular belief, the bear cub is born as a shapeless and eyeless lump of flesh. The mother bear gives it form by licking it: *quod ore suo formet fetus* (Fig. 5.3). This etymology comes from Isidorus, *Etymologiae* (Etymologies II, 22), despite the very roots of this belief being much older. Some traces of this idea can also be seen in Ovidius' *Metamorphoses*:

> *Nec catulus, partu quem reddidit ursa recenti, Sed male viva caro est; lambendo mater in artus Fingit et in formam, quantam capit ipsa, reducit.* (Ovidius XV, 379)
>
> The cub that a mother bear has just given birth to is not a cub but a barely living mass of flesh: by licking it the mother gives it a shape and gives it the form she has herself.

A very similar description was provided by Plinius Maior in *Naturalis Historia*:

> *Hi* [the bear cubs] *sunt candida informisque caro, paulo muribus maior, sine oculis, sine pilo; ungues tantum prominent. Haec lambendo paulatim figurant.* (Plinius VIII, 54)
>
> When first born, they are shapeless masses of white flesh, a little larger than mice; their claws alone being prominent. The mother then licks them gradually into proper shape. (Translation: H. Rackham 1940, 91)

We can also see that, besides Isidorus, another *auctoritas*, the Persian peripatetic philosopher Avicenna (Ibn Sīnā), is also referred to. In any

Fig. 5.3 Bear licking her cub into a form. Source: Den Haag, Huis van het boek, 10 B 25, folio 11v. With permission of Huis van het boek

case, it is necessary to verify the provenience of quotations as the notion of the bear moves from one medieval authority to another. This is true for every work, but it becomes essential in relation to medieval encyclopedias, since, as previously remarked, one of their typical features is to declare sources which are not always the authentic ones. This means that one may end up being unable to find the original quotation itself. In other words, quotations in medieval encyclopedias quite frequently do not derive from the source that was cited by the author as original, but rather from an intermediary one. In this case, the information reported as having been written by Avicenna occurs in the work of Isidorus and in Bartholomaeus Anglicus as well.

As for Isidorus, the passage can be found in the already mentioned *Etymologiae*:

> *Ursus fertur dictus quod ore suo formet fetus* [...]. *Nam aiunt eos informes generare partus, et carnem quandam nasci quam mater lambendo in membra componit.* (Isidorus II, 22)
>
> The bear is called this way because it forms the cub with its mouth—*ore suo*. In fact, she-bears are said to deliver shapeless cubs and to form their bodies by licking them once they are born.

At the same time, this is the text from *De proprietatibus rerum* by Bartholomaeus Anglicus (ed. Nürnberg 1492, F [6r]):

> *Ursus est vocatus eo quod ore suo formet fetus* [...] *ut dicit Isidorus libro XII. Nam aiunt eos informes generare partus et carnem quandam nasci quam mater lambendo membra componit.* (XVIII, 110)
>
> The bear is called this way since it forms its cub with its mouth, as Isidorus affirms [...]. In fact, she-bears are said to deliver shapeless cubs and to form their bodies by licking them once they are born.

To date I have found no evidence of a similar passage in the corpus of the *Avicenna latinus*. Therefore, it is likely that Giovanni took these traits either from one of the aforementioned sources or from another intermediary not as yet identifiable.

Conclusions

In this chapter, I tried to analyze the way Giovanni da San Gimignano used zoological material to build the structure of a typical *similitudo* and of an *exemplum* in *LdE*, which were intended to illustrate what constitutes the characteristics of a good preacher, moving animals between past authorities and from the natural world onto the constructed world of the written word. It could thus, be elucidated what may be hidden behind the not always "reliable" sources cited in medieval encyclopedias. Such discrepancies should not be a surprise, since Humanism was still far in the future and the approach to these texts has not engendered much philological attention. Future research will hopefully reveal new elements as well as suggest new perspectives in the study of animals in the *LdE*.

In this concise study I intended to briefly focus on how Giovanni da San Gimignano's work specifically functioned and the way the author used *similitudines*, in particular those which are drawn from the animal world, in order to build a preaching instrument that was widely diffused across Central and Eastern Europe and which was to become a successful aid for preachers throughout the centuries to come.

References

Primary Sources

Bartholomaeus Anglicus. 1492. *De proprietatibus rerum*, Nuremberg (Bavière, Allemagne): Anton Koberger.

S. Bonaventura. 1934. *Collectiones in Hexaëmeron,* edited by Fernando Delorne, Firenze: Quaracchi.

Giovanni da San Gimignano, *Liber de exemplis et similitudinibus rerum*, Anvers 1609.

Gregorius Magnus. 1999. *Homiliae in evangelia*, edited by Raymond Etaix. Turnhout: Brepols.

Isidorus Hispalensis. 1911. *Etymologiarum siue Originum libri XX*, 2 vol., edited by Wallace M. Lindsay. Oxford: Oxford University Press.

Ovidius. 1981. *Metamorphoses*, edited by William S. Anderson. Lipsiae: Bibliotheca Teubneriana.

Plinius Maior. 1892–1909. *Naturalis Historia*, 5 vol., edited by Ludovic Ian, and Carolus Mayhoff. Lipsiae: Bibliotheca Teubneriana.

Plinius Maior. 1940. *Naturalis Historia* 2nd vol., edited and translated by Harris Rackham. London–Cambridge MA: William Heinemann–Harvard University Press.

Thomas Aquinas. 1953. *Catena aurea in quattuor Evangelia* vol. V, edited by Angelico Guarienti. Torino: Marietti.

Secondary Sources

Boutell, Mary E. C. 1869. *Picture Natural History.* Chicago, Ill.: Thompson and Thomas.

Dondaine, Antoine. “La vie et les œuvres de Jean de San Gimignano.” *Archivum fratrum praedicatorum* 9 (1939). 128–183.

Kaeppeli, Thomas. 1975. *Scriptores Ordinis praedicatorum Medii Aevi*, 4 voll., Romae 1970–1993, II, 1975. Romae: Panella, 539–543.

Kaeppeli, Thomas. 1993. *Scriptores Ordinis praedicatorum Medii Aevi*, 4 voll., Romae 1970–1993, iv, 1993. Romae: Panella, 170.

Ledda, Giuseppe. "Filosofia e ottica nella predicazione medieval." In *Letteratura in forma di sermone. I rapporti tra predicazione e letteratura nei secoli XIII–XVI*, Atti del Seminario di studi (Bologna 15–17 Novembre 2001), edited by Ginetta Auzzas, Giovanni Baffetti, and Carlo Delcorno, 53–78. Firenze: Olschki, 2003.

Oldoni, Massimo. "Giovanni da San Gimignano". In *L'enciclopedismo medieval* (Atti del convegno San Gimignano, 8–10 ottobre 1992), edited by Michelangelo Picone, 213–228. Ravenna: Longo, 1994.

Quétif, Jacques, and Jacques Échard. 1719. *Scriptores Ordinis praedicatorum.* Paris: Ballard et Simart.

Raffin, Pierre. "Jean de San Gimignano." In *Dictionnaire de spiritualité* VIII, edited by Marcel Viller, Charles Baumgartner, and André Rayez, 721. Paris: Beauchesne, 1974.

Stegmüller, Friedrich. 1951. *Repertorium biblicum Medii Aevi* vol. III, Matriti: Instituto Francisco Suárez.

Van den Abeele, Baudouin. "Moralisierte Enzyklopädien in der Nachfolge von Bartholomaeus Anglicus: das 'Multifarium' in Wolfenbüttel und der 'Liber de exemplis et similitudinibus rerum' des Johannes de Sancto Geminiano." In *Die Enzyklopädie im Wandel vom Hochmittelalter zur frühen Neuzeit*, edited by Christel Meier-Staubach, 279–304. München: Fink, 2002.

Vecchio, Silvana. 2001. "Giovanni da San Gimignano." In *Dizionario biografico degli Italiani*, edited by Mario Caravale, 206–210. Roma: Istituto dell'Enciclopedia italiana.

Von Moos, Peter. "L'exemplum et les exempla des prêcheurs." In *Les exempla médièvaux: Nouvelles Perspective*, edited by Jacques Berlioz, and Marie Anne Polo de Beaulieu, 67–82. Paris: H. Champion, 1992.

Zambon, Francesco. 2001. *L'alfabeto simbolico degli animali.* Milano–Trento: Luni Editrice.

CHAPTER 6

Always Angular and Never Straight: Medieval Snakes in Human Graves?

Monika Milosavljević

Introduction

Snake skeletons are generally uncommon in archeological contexts, let alone burials. Although several snake skeletons were found between 1994 and 1996 in seven separate graves at the Roman and medieval cemetery of Ravna–Slog (Eastern Serbia), they received little attention at the time as they were regarded as being a naturally occurring phenomenon. This interpretation, however, is not congruent with the behavior of snakes at death. Notwithstanding, the placement of the snake skeletons in a cultural deposit raises more questions than it answers.

While a single occurrence of a snake being discovered in a grave may be attributable to natural causes, several such discoveries from a series of graves suggest that these are not random occurrences, but rather deliberate placements. Unfortunately, due to decisions made by the archeologists in the field, the snake skeletons were later discarded. The skeletons were documented in accordance with standard practice at the time, but not

M. Milosavljević (✉)
Faculty of Philosophy, Department of Archaeology, University of Belgrade, Belgrade, Serbia

L. Bartosiewicz, A. M. Choyke (eds.), *Medieval Animals on the Move*, https://doi.org/10.1007/978-3-030-63888-7_6

photographed. However, thanks to the extant documentation, records do exist of their connection to the human remains, indicating that the snakes and the deceased were buried together. While these drawings remain the only information on which this case study can be based, and regrettably are insufficient for detailed analyses, the phenomenon itself is of epistemological relevance for the study of human–animal relations in archaeology (Petković et al. 2005; Špehar 2017, 114–117).

Materials and Methods

The Ravna–Slog site in present-day eastern Serbia is located in the vicinity of the Roman *castrum* of *Timacum Minus* (AD first to sixth centuries), which subsequently became an early Byzantine fortress. Excavations at this site in the 1990s uncovered 72 late Roman graves from the fourth to fifth centuries. These graves contained 75 individuals. In addition, 65 medieval graves (AD ninth to eleventh centuries), containing the same number of skeletons, were found in this (Petković et al. 2005; Ilijić 2015, 24–34).

Although it has been assumed that there was a cultural hiatus between the late Roman and early medieval burial strata despite common practices in early medieval burial customs, excavations have shown that the late Roman cultural space was re-used during the Early Middle Ages. The early medieval funerary rite included inhumation of the deceased in simple rectangular grave-pits dug into the existing Roman *memoria* (masonry tombs) throughout the Roman cemetery. The East–West-oriented burials were arranged in parallel rows with over-lapping graves but were not conclusively Christian. The deceased were put into the grave in a supine position. Only one early medieval grave (no. 61) dug into a Roman *memoria* was oriented North–South, with the head pointing South. Offerings were found in 40 graves and included garments and decorative objects (e.g., earrings, rings, necklaces, buttons, and pendants), weapons (e.g., knives, axes, arrow heads), and pots. The best represented jewelry category was earrings (80) found in 22 graves (Petković et al. 2005, 181–243; Ćirić 2014, 108–110).

According to the physical anthropological analyses of the published graves, the 12 males and 20 females belong to the early medieval population. Of the 33 skeletons whose sex could not be identified, 29 were children, mainly infants. Skeletal remains of snakes were found in five of these early medieval graves: four separate graves (no. 11, 24, 73, 120) contained female skeletons and one snake came from a child's burial (no. 37). Snake skeletal remains were also found in other graves supposedly dating to the Roman period: in one female (no. 4) and one child's grave (no. 53). However, the dating of these graves is open to debate due to the re-use of the grave site disturbing the Roman stratum and the probable reuse of objects representing Roman material culture. The medieval component of the Ravna–Slog cemetery has a higher proportion of infant burials compared to Roman graves at the site. Additionally, the high incidence of blood and metabolic disorders as well as evidence for reconstructed neglected skeletal trauma indicate that early medieval living conditions were dire, with poor medical practice and malnutrition (Petković et al. 2005; Petković et al. *in press/* 2015b).

As per the original site documentation, a reconstruction of the snakes' positions suggests that their placement together with the human corpses was deliberate. The three (medieval) female skeletons had snakes wrapped around their waists like "belts." In two cases, the tails of the snakes extended up either the left or right shoulder of the deceased. Moreover, the child graves from both the Roman and early medieval periods, contained skeletons of snakes laid straight on either the right or the left side of the supine body (Fig. 6.1). This clearly indicates that the snakes were used in the funerary practice during the interment ritual (Petković et al. 2016, 89–94).

The cultural deposit theory would complicate matters since the same practice was apparently being used in both Roman and early medieval times in the same cemetery (Ilijić 2015, 29; Petković et al. 2016, 32–35; Petković and Miladinović-Radmilović 2013, 47–60). Such continuity in practice would be hard to account for given the long timespan and the probable cultural shifts in burial practice by the local population. The dating itself, however, was based only on a low-resolution typochronological analysis of finds. In the detailed report of the 1994–1996 excavations (Petković et al. 2005), 120 individuals were assigned to two cultural strata, late Roman (AD fourth to fifth centuries) and early medieval (AD ninth to eleventh centuries). Small-scale excavations conducted in 2013–2015 have not been systematically published. They brought to light 50 late Roman and early medieval graves (Petković et al. *in press*/2015b). Several of the graves date from the eighth century AD onward, that is, an earlier medieval burial horizon.

Within the cemetery's dating to the ninth to eleventh centuries, the Ravna–Slog graves containing snake skeletons date to the end of the ninth to tenth centuries based on the typochronology of earrings found in graves no. 11 and 73 (Fig. 6.2). Absolute (radiocarbon) dates might help in

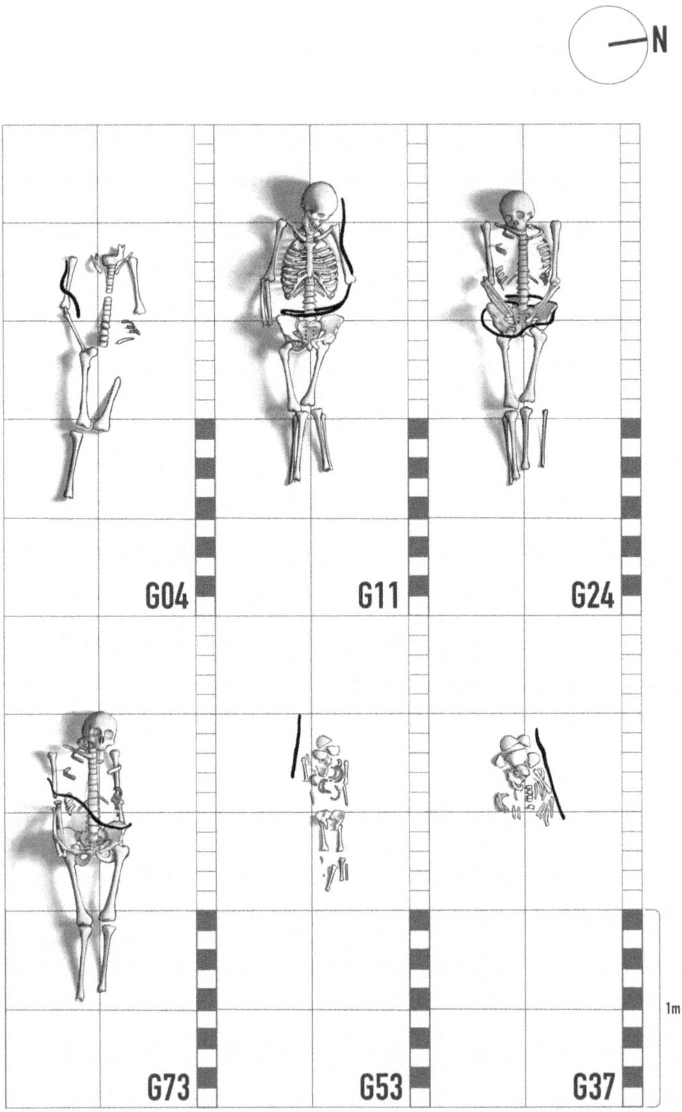

Fig. 6.1 The placement of the snake skeletons (graves no. 4, 11, 24, 73, 53, 37) at Ravna–Slog. Visualization by Nikola Stepković

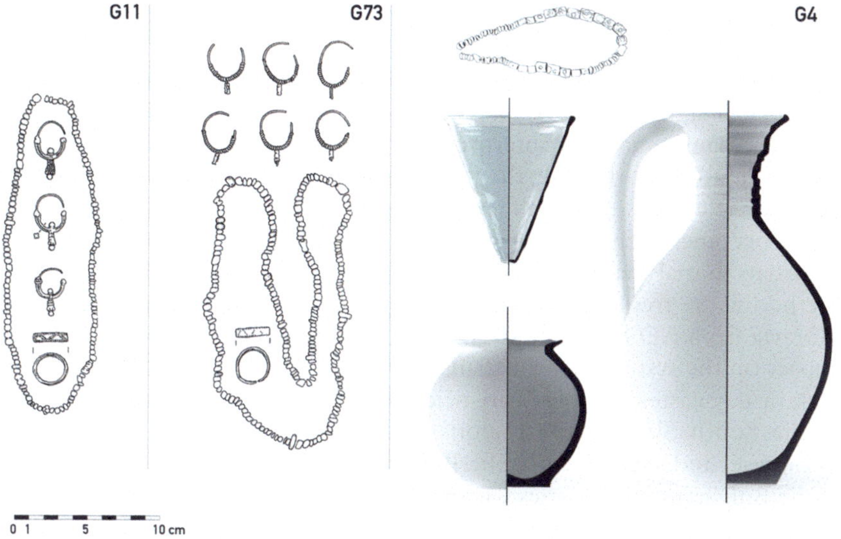

Fig. 6.2 Archaeological artifacts associated with snake remains (graves no. 4, 11, 73). Visualization by Nikola Stepković

determining a proper date, but these have yet to be carried out. Notwithstanding, ^{14}C dating might also be inaccurate due to a so-called plateau in the calibration curve for this time period.

The territory of Ravna–Slog changed hands multiple times during its history. While it is well known which major political power held sway over the ninth-to-tenth-century settlements in its vicinity, these settlements still lay outside the reach of Byzantine administration and military power until the eleventh to twelfth centuries. It was not until that time that Byzantium established a short-lived northern frontier along the Danube for its Western Balkan territories. Given this ephemeral shift in "Empires," the Ravna–Slog settlement probably still lay on the periphery of the Bulgarian Empire when the early medieval burials took place. The cultural heterogeneity inherent to the region at the time, compounded by the loose association of state structures for early medieval empires (particularly the medieval Bulgarian Empire), means that the burials do not reflect the cultural norms of a state. Christian burials at the time of conversion primarily represented elites as early adopters of social innovation and political expedience. Christian burial practice is less manifest among the lower social strata resulting in a greater heterogeneity of burial rites (Ostrogorsky 1968, 210–298; Marjanović-Vujović 1997, 236–246).

Burying humans with animals or animal parts at this site has been interpreted as a sign of local pagan practice. As such, any graves containing animal remains tend to be interpreted as coming from pre-Christian times. Burial datable to the fourth and fifth centuries might easily contain Christian elements, thus, the vague distinction between Christian and pagan becomes even fuzzier. However, it is unclear when a shift to stricter Christian burial habits occurred in this region. Although Cyril and Methodius are frequently cited as having introduced Christianity to Great Moravia from AD 867 onward, it is not clear whether their work had any impact on the systematic Christian conversion of villagers in the region of present-day eastern Serbia (Ferjančić 2009, 47–53). The Christianization of the South Slavs in the region occurred under the influence of the Byzantine, and subsequently, the Bulgarian Empires. Under the influence of the Bulgarian missionaries Clement (AD 840–916) and Naum of Ohrid (AD 830–910), followers of Cyril and Methodius, the South Slav conversion became more marked. However, it cannot be asserted that this process was complete and that all cultural activities practiced prior to the adoption of Christianity were abandoned (Špehar 2017, 101–105). While Christianization may be seen as the *terminus ante quem* for human burials with animals in the early medieval context, there are important exceptions. In grave no. 226 of an adult female at the site Mačvanska Mitrovica–Zidine, the grave goods included a bronze "encolpion" cross (reliquary) dated to the eleventh century, one perforated animal tooth, and one copper coin (Ae3) of the Roman emperor Valentinian I (AD 364–375). The skeleton was interred in a supine position with the arms lying on the pelvis and a Roman brick placed beneath the head (Ercegović-Pavlović 1980, 22; Ćirić 2014, 55–56). Based on a detailed analysis of commoners' cemeteries, Uwe Fiedler has identified a number of changes in burial practice attributable to Christian conversion: 1) change in orientation from unspecific to East–West, 2) disappearance of cremation; 3) disappearance of grave offerings (animal bones and eggs); and 4) increase in the deposition of dress accessories in female graves (Curta 2018, 257). In Merovingian archaeology, a shift from burial with goods to the significance of the grave's location relative to the church is regarded as indicative of the active presence of Christianity (Effros 2003, 171–173). Multiple grave sites that fall short of criteria for a "Christian" burial remind us that caution should be used when attempting to establish clear-cut lines of past social change

(Džino 2010, 150–154). Such shifts are often interpreted as instantaneous, uniform, and concurrent in hindsight as opposed to gradual but expeditious change. Moreover, not all Christian practices are mutually exclusive from pagan tradition. Paying attention to animals and burying them alone or with humans, as well as putting meat (partially as sacrifice or other ritual purposes) in human graves, was not totally absent from early medieval Christian customs. The practice of including animals in graves ceases by the onset of the tenth century in the Central Balkans (Špehar 2017, 105). Bearing this in mind, the presence of snake remains in the graves at Ravna–Slog is not a precise criterion for relative dating.

Medieval women buried with snakes are of particular importance at the Ravna–Slog site (Table 6.1). The buried women do not receive the same kind of burial goods nor were they the same age at death. Several of the deceased were buried with jewelry (a ring, necklaces made of yellow, blue, and white glass beads, as well as bronze earrings). Other graves without snake remains also contained significantly more luxury jewelry in terms of

Table 6.1 Graves containing snake skeletons at Ravna–Slog. Ages are given in years

Grave	*Age*	*Sex*	*Assorted grave goods*	*Position of snake*
Medieval, no. 11 (1994)	16–18	F	Six bronze earrings (ninth to tenth centuries), necklace of 226 glass beads (yellow, blue, white), one ring	- Wrapped as a belt, extending up to the left shoulder
Medieval, no. 24 (1994)	16–18	F	No finds apart from the snake remains	- Wrapped as a belt
Medieval, no. 37 (1994)	3	/	No finds apart from the snake remains	- Elongated along the left side
Medieval, no.73 (1995)	42–48	F	Three bronze earrings (tenth century), a necklace of 146 glass beads (yellow, blue, white), one ring, one pot containing eggshells	- Extending from the pelvic region to the right shoulder
Medieval, no. 120 (1996)	44–50	F	Iron scraper, Lat. scalprum	- Unknown
Roman (?), no. 4 (1994)	46–54	F	One necklace made of glass beads, one conical glass goblet, one small kilned beaker, and one kilned flagon	- Extending from the pelvic region up to the right shoulder
Roman (?), no. 53 (1994)	3–4	/	Eggshell	- Elongated along the right side

both raw material and manufacture. Small pots and/or eggshells were also buried in graves where snake remains were found. Consequently, the female and child dead interred with snakes do not differ from other graves in this cemetery in terms of any signifiers in the grave offerings they received. This fact necessitates a further examination of the site.

Archaeological data may be seen as interpretative constructions. Due to their role in evidential reasoning (rather than having an exclusively empirical content) data have a privileged status in research. They may, however, be biased by the perception of the individual collecting them. When regarded as unimportant to record, certain aspects of data may be lost for future analyses (Babić 2015, 256–258; Chapman and Wylie 2016, 93–136; Milosavljević 2016, 96–97). Therefore, data deserve re-examination when prudent and promising:

> The literal retrieval of and source criticism of primary data, both records and material finds, and the reanalysis by which secondary claims about their interrelationships are apprised can significantly reconfigure the 'facts' that ground evidential claims. This process of critically scrutinizing the facts can, in turn, destabilize not just the interpretative and explanatory hypotheses they had been recruited to support as evidence, but also the scaffolding of assumptions that had informed their capture and systematization in the first place. (Chapman and Wylie 2016, 100–101)

Accordingly, animal-free (in this case, snake-free) contexts in early medieval archaeology in Serbia do not always mean the absence of animals from the archaeological record, but rather mean that animal remains were not treated as relevant to archaeological research. These professional practices have changed in Serbia since 2000 as zooarchaeology (Živaljević 2013, 1137–1163) and theoretical archaeology (Milosavljević 2016, 90–91) have garnered more academic attention at the Department of Archaeology, Faculty of Philosophy, University of Belgrade.

Data gathering at Ravna–Slog was based on an outdated paradigm. The analysis, thus, has proven difficult since most data originate from old excavations and funding for new and original research on such materials is not easily available. The only remaining strategy is the meticulous re-examination of the archaeological records still available. The analysis of materials from Ravna–Slog was carried out as follows:

A. If the snake remains can be conclusively identified as originating from natural deposits, then the Ravna–Slog may be interpreted as a typical early Slavic cemetery similar to other cemeteries in the same region.
B. If the snake remains turn out to originate from cultural deposits, it would then be unclear why snakes were added to both the two Roman and five medieval inhumations given the chronological differences. There could be two mutually exclusive possibilities: (1) a continuity in culturally transferred ritual practice or (2) the typo-chronological misinterpretation of the two Roman graves, as there are notable instances of the reuse of Roman artifacts in early medieval contexts at this cemetery. Although there are respected authors who have claimed a continuity of practices from the late Roman period to subsequent ethnically defined populations (even up to the present), option B is the more plausible if the snakes can indeed be considered cultural deposits.
C. If the snake skeletons can be safely identified as cultural deposits *sui generis* for the Early Middle Ages, then the phenomenon of reusing Roman material culture should be discussed in light of burying snakes with the deceased at Ravna–Slog. Archaeological analogies for human inhumation accompanied by animals (specifically snakes or snake parts) and recycled Antique objects from the original Roman cultural space, provide additional support.

In order to establish whether the snake remains were natural or cultural deposits, a herpetologically based knowledge of snake behavior and taphonomic insight are crucial (Albarella et al. 2017, 769–770). In order to properly understand why the snake skeletons were neglected in previous archeological interpretations, the sociology of knowledge needs to be understood. Conforming to Ludwik Fleck, a collective body—for example, a scientific community—intrudes into the process of looking and seeing, thinking, and cognition (Fleck 1986 [1947], 137–140). Consequently, archaeological recording is based on ecologies of practice, such as disciplinary culture and tradition, the training and funding of the system used, as well as institutional infrastructure. All these bolster or upset actual practice. Archaeologists of the time, through no fault of their own, likely did not consider the snake skeletons unique, worthy of detailed documentation and collection. This article seeks to readdress findings made at the excavation. It also draws upon the new context of archaeological excavations and documentation that have become more common in Serbia since 2000, improving the quality of data in terms of complexity and depth. Although recent excavations at Ravna–Slog have not uncovered new skeletal remains of snakes, the quality of data on re-using material culture, including intentionally deposited animal remains, aids further reinterpretation, even if the supposed practice of entombing humans with snakes is problematic.

Studying animals in the human past has recently become a trend in anthropology, archaeology, history and the history of art, as part of the overall "animal turn" (Ritvo 2007; Cederholm et al. 2014). Critical animal as well as human–animal studies from the domain of socio-cultural anthropology can be integrated with medieval archaeology in order to broaden traditional medieval topics and add to their understanding. Animals must be considered beyond livestock breeding, hunting, and the evidence of exotic animals: their presence also provides insight into the lost intricacies of daily life in the past.

While animal remains in graves are easily interpreted when ascribed to a consumable resource or to a totem, it is not as easy to designate them outside these two rough categories (Pluskowski 2007; Hill 2013, 117–136). Unfortunately, however, the Ravna–Slog snake remains fit neither of them. They do not occur singularly, however, and may be seen in light of the reuse of artifacts with which they were buried. Evaluating them within the context of reuse may reveal the original intent behind their deposition.

Medieval archaeology must incorporate a multitude of perspectives in order to better uncover the past. Theories concerning the "biography of things" and cultural memory (Kopitoff 1986; Asman 2011; Assmann 2011) show that the meaning of material culture is fluid and can change along with the social context. The phenomenon of reuse therefore can serve as a referential framework when analyzing Antique material goods found in early medieval contexts. Following shifts in the fluidity of usage may also reflect social shifts during the Early Middle Ages of the Balkans.

Discussion

The initial question is whether snake remains from the cemetery should be considered natural deposits. If the snakes in the Ravna–Slog graves died accidentally, they would be collateral to the burials and unimportant for further interpretation. However, if they occur as a result of human intent, the opposite would be true.

According to the herpetologist, Rasko Ajtić (personal communication, January 25, 2016), based on the lengths of the snakes found, the skeletons probably belonged to one of two non-poisonous species: the four-lined snake (*Elaphe quatuorlineata)* or Caspian whipsnake (*Dolichophis caspius*). The four-lined snake is not native to eastern Serbia since it requires a Mediterranean climate. Its behavior is generally more docile than that of other snakes which would have allowed them to be safely transported from their original habitat. In contrast, the Caspian whipsnake is native to the entire Balkan region. While it is not venomous, it can be aggressive (Đorđević and Tomović 2015, 213–219; Tomović and Džukić 2015, 220–226).

If snakes indeed form part of cultural deposits, their use also depends on their seasonal availability. Serbian ethnologist Tihomir Đorđević (albeit relying on unquestioned traditional views) noted that there were many belief-driven customs in the Balkans in the mid-twentieth century in which snakes played a prominent role. Reportedly, awakening from hibernation they were commonly greeted with singing by adolescent girls to encourage the snakes' reproductive cycle. Touching "dried snakes" had medicinal powers. Snake heads were recorded as having a garlic clove stuffed into the mouth as a seed to plant and thereafter being used as an amulet. These examples are not direct ethnographic analogies, but illustrations of the diverse, possibly seasonal roles, snakes played in culture (Đorđević 1958, 100–186).

There have been recent re-interpretations of the Ravna–Slog graves with snake remains to suggest the evidence of Dionysian cult activities (Petković et al. 2016). Essential to this re-interpretation is putative cultural continuity between the Roman and medieval practices at the site as well as ethnographically recorded Serbian folk tradition. The archaeological phenomenon could be thus explained by the conservatism in burial practices, implying uninterrupted cultural continuity invisible in the archaeological record. Reference to Asclepius, as represented by snakes and their healing power, remains problematic in this argument until a direct link can be demonstrated between the archaeological and ethnographic evidence.

Biological Arguments

Evidence on snake behavior, especially during hibernation, is of crucial importance here (Bartosiewicz et al. 2013, 77–83, Lenders and Janssen 2014, 319–346; Tomović et al. 2014, 129–158). Intrusive animals are regularly encountered in graves during field work. Through careful excavation one can usually identify animal burrows in the burial. When an entire snake skeleton is present, it may be a "dead give-away" as an intrusive entry. Unable to dig, snakes use burrows of small mammals (e.g., hamsters or rabbits) and naturally occurring crevasses (Tomović and Lakušić 2015, 49–63). Unfortunately, all the data on the graves at Ravna-Slog contain no record of any animal burrows or evidence thereof found during excavation.

Nonetheless, the behavior of snakes may explain their presence in the graves. Given their hibernation habits, death near or inside a grave is a distinct likelihood. Snakes can be attracted by the man-made "burrows" created by funerary activities, penetrating through such cracks and cavities into the grave. Temperatures turning too cold may prevent the snake recovering from hibernation (Reitz and Wing 2008, 90).

Snakes often hibernate in large groups, sometimes including various reptile species and amphibians. Animal burrows, however, rarely go below 0.5 m (*personal communication* with László Bartosiewicz, ZOOARCH mailing list January 8, 2016). Most human burials at the Ravna–Slog cemetery were recorded at a relative depth between 0.5–2 m at the time of excavation, largely excluding such shallow natural snake deposits (Petković et al. 2005). Temporarily open grave pits as natural traps may be another source of natural deposition (Bartosiewicz et al. 2013, 82).

Snakes do not truly hibernate but exist in a state of reduced activity (brumation), which means they remain responsive to their surroundings. During brumation, they coil or intertwine with others in order to minimize heat loss. If weather conditions (such as flooding or extreme temperatures) affect the burrow, snakes are still able to react. They die of starvation only during the harshest winters. Snakes can also escape from forest or moor fires in their hibernaculum. Death during brumation is most commonly attributable to sudden rises in groundwater (Todd et al. 2009, 1221–1222).

Given these behavioral traits, it would be rare to repeatedly find singular snake skeletons in several graves but more common to find clusters of snakes in few, exceptional deposits. The situation at Ravna–Slog is just the opposite: remains of several snakes were found singularly and not coiled. If complete individual skeletons of snakes are found in closed human graves (and not coiled or in a tangle with other snakes), it is highly likely that they were deposited there at the time of the burial. Moreover, dead snakes would be unable to assume a straight position or a position in which their elongated body would be coiled around another object. Instead, it seems likely that at Ravna–Slog they were stretched out or placed around or near the human body deliberately.

Archaeological Analogies

Considering that snake remains are rarely found in either natural or structured archaeological deposits, every analogy can be useful regardless of its cultural or chronological context. The best analyzed example of snake remains was published from an Early Copper Age grave at Pusztataskony–Ledence I, Hungary (Bartosiewicz et al. 2013). That paper details the environmental and cultural implications of non-human vertebrate skeletons from a prehistoric inhumation in east-central Hungary in which, among others, a dozen non-venomous snakes (representing three species) were recovered alongside and underneath a woman's skeleton (Bartosiewicz et al. 2013, 77–88). Of 39 Dilmun Period, Iron Age, deposits in Qualat al-Bahrain, 32 contained complete snake skeletons. The vast majority, described as "sacrificial," was found in ceramic vessels, set into pits in the floor (Potts 2007, 55–74). Regardless of their distance from Ravna–Slog in time and space, these examples show that deliberately placed snakes may form an integral part of the archaeological record.

Culturally relevant snake remains have been recovered in Hungary and found only in burial contexts dated to the period of the Avar Khaganate (ca AD 580–800).

At the site of Vác–Kavicsbánya, grave, no. 366 of a male was robbed, leaving its contents in disorder. However, a small, hand-thrown pot, containing 11 snake vertebrae, was found in its original position. In addition, an iron fibula, a light grey, wheel-thrown pot and two eggs were found within this tomb, dated to the early Avar period (Szentpéteri 1993, 226).

Ten vertebrae of an outstretched snake were recovered from grave no. 14 of an adult woman, dated to the late Avar Period cemetery of Závod in south-western Hungary. Approximately 200 beads of different shapes and sizes were also found in this grave, including ten snake vertebrae, one colored platinum-green. A large iron chain consisting of different-sized links was discovered between the left and right upper arms. The smallest link was on the left and the largest on the right upper arm. Ten small links (1 cm in diameter) were connected to each other. Between the links of the chain, one tubular object and half of a bronze button were also found (Kiss 1984, 161–185).

At the Avar cemetery of Gyönk–Vásártér, the stretched skeleton of a decapitated snake was found in grave no. 227 next to the right hand of an adult female. The burial contained a fragment of a Roman crossbow-shaped brooch above the skull, a bronze earring on the right side, a black wheel-thrown pot alongside the right femur and, fragmented faunal remains. The front section of the snake was placed near the woman's shoulder while the rear section was wrapped around the pot. The excavation report claims this to be a unique example, adding that the grave bore no evidence of having been opened prior to excavation. It also asserts that the headless snake is proof that this animal had been buried alongside the human. Although no closer date has been established, this grave is also from the period of the Avar Khaganate (Rosner 1972, 210–211; Szentpéteri 1993, 226).

The three Avar examples are culturally and chronologically closest to the medieval contexts at the Ravna–Slog cemetery. The similarity may not be simply seen as direct cultural interaction. It is possibly attributable to an "exceptional" knowledge transfer from generation to generation. The Avar Khaganate existed in Central Europe between the sixth and eighth centuries (Pohl 2003, 571–595), while the site of Ravna–Slog is dated to a time period following immediately after its existence, lasting until the beginning of the eleventh century. It is possible although not certain that the deceased women were guardians of specific Avar traditions and practices. In the aforementioned Avar examples, however, there was only one person buried with a snake in each cemetery, whereas at Ravna–Slog several cases were discovered. An additional argument for an Avar connection is that eggshells are frequently discovered in graves at Avar cemeteries. Eggs are commonly interpreted as symbols of fertility and rebirth (Premužić et al. 2013, 59).

An Early Medieval Phenomenon

Traditional interpretations of objects recovered from graves predominantly rely on typological analyses within a cultural-historical explanatory model: artifacts tend to be taken as tokens of the ethnicity or religion of the deceased. Current research redirects the focus onto objects as signifiers of social status and local identity within the burial community. One way social changes can be compared between the Balkans and Western Europe (Härke 2014, 41–60; Džino 2014, 127–152) is through differences in the typochronology of particular artifacts used to denote established demarcations of cultures.

Post-Roman, post-imperial cemeteries in Western Europe are also reasonable parallels for early medieval cemeteries in the Balkans. In these contexts, social status was displayed by choosing the place of a burial site and/or by placing particular objects in graves although this was not typical for all burials. Such burial assemblages raise the question of "mixed-artifact contexts" (Roman and medieval) at the Ravna–Slog cemetery. Archeological reports have treated these as anomalous since their alternative explanations do not fit standard interpretation. This approach, however fails to address the uniqueness of these graves. Their exceptionality should be treated as a possible sign of the way the local community represented the social identity of the deceased at the moment of burial.

The re-examination of Roman graves containing snakes with regard to early medieval dating is pertinent here. Additional mixed-burial contexts characterized by typochronologically mixed finds from both the Roman and early medieval periods were discovered at Ravna–Slog during the 2013–2015 excavations. These data are more detailed than previous evidence from the 1990s (Petković et al. *in press* /).

"Foxy lady," grave no. 159, was found to contain skeletal fox remains wrapped around the neck of a female. The fox skeleton was found on top of a bi-conical Roman flagon. The woman was buried with a bronze ring, a knife, necklaces made of glass beads and seven bronze earrings—all representing early medieval material culture. This admixture poses clear evidence of an intentionally composed assemblage at the time of burial using grave goods comprising Roman period as well as contemporaneous objects (Petković et al. 2015a, 89–90).

According to the physical anthropological analysis, the human skeleton belongs to an adult female, aged 40–45 years. Degenerative lesions on her skeleton resulted from osteoarthritis. Markers of occupational stress (hypertrophy on muscular surfaces) are a clear indication of heavy physical activity throughout her life. There were few teeth found in the maxilla, but all mandibular teeth were lost *ante-mortem*. This status could be interpreted as periodontal disease (not remarked in the maxilla) or as a result of mutilation. She was laid into an oval grave-pit in a supine position with the arms extended alongside her body. The human skeleton is poorly preserved and the skull is fragmented. The northwest section of the grave was disturbed by subsequent digging activity, which partially destroyed the left side of the human and part of the fox skeleton. The left arm bone is missing while bones of the right forearm were displaced, although the hand was found intact *in situ*. Bones of the left side of the chest are also missing or dislocated (Petković et al. *in press*).

The fox skeleton is almost completely preserved. The animal was wrapped around the woman's neck, lying on its left side with the head oriented north. It was likely a fully grown male (aged 2–4 years). There are no traces to indicate the cause of death, but judging from the condition of the skeleton, the fox was buried shortly after its death before *rigor mortis* set in (Petković et al. 2015a, 89–90; Petković et al. *in press*/2015b). This unique context is analogous to a case in the "Bečej–Pionirska Street" cemetery in northern Serbia, attributed to the First Avar Khaganate (dated before AD 626): at Bečej a disturbed grave (no. 16) contained the skeleton of an adult woman and the complete skeleton of a fox (Mikić-Antonić 2012, 14–15).

Grave no. 142 at Ravna–Slog contained the skeleton of a child around seven years old, buried in a supine position. The grave pit was covered by large pebbles and other roughly broken stones. The deceased had a set of tools near the left hand: an antler-handled spear drill, an iron chisel and three unidentifiable iron objects filled with lead. One rough, hand-thrown pot with "Early Slavic" decoration was also found, as well as a Roman flagon. This "mixed context" was highlighted in the excavation report as an enigma (Petković et al. 2014, 79–80).

Graves no. 4 and no. 53 putatively date to the Roman period. Both contain the remains of a snake. Grave no. 4 of an adult woman was rich in grave goods including a glass bead necklace, a conical glass goblet, a small fired clay beaker, and a fired clay kilned flagon. Snake remains were found along the right arm. Grave no. 53 contained eggshells but has no other artifacts and thus, cannot be reliably dated (Petković et al. 2005, 25, 34; Petković et al. 2016, 19–20). Since these graves can only be dated through their material culture, it is possible that artifacts found in grave no. 4 (dated as Roman on a typological basis) were actually Roman artifacts reused in medieval times. Such an assertion would be more likely considering the snake skeletons which would not in themselves be of conclusive dating value without additional supporting evidence.

Apart from these unique graves, there are also contexts containing reused artifacts at Ravna–Slog, interpreted as medieval despite the evidently mixed typochronology of their find materials. Grave no. 98 contained a woman (29–35 years), buried with a necklace composed of 99 glass beads and a reused perforated Roman coin (not readable). There were also six bronze earrings, two bronze rings, and an iron knife found in her grave. Furthermore, medieval grave no. 47 containing a young woman (17–18 years), yielded a necklace with a prehistoric arrowhead strung on it reused as a pendant (Petković et al. 2005, 187, 194; Ilijić 2015, 74–75).

Comparing the reuse of objects from earlier periods to the presence of snake skeletons, the dating of the graves is not in accordance with Occam's razor. It would simplify the entire issue if the material culture was seen as reused and not inherent to its original use, so that all graves containing snake remains could be dated as medieval. This supposition would not be conclusive and does not account for inherited ritual and/or shared traditions despite the long period involved between the late Roman and early medieval cemeteries. However, the conclusion would be congruent in terms of mixed assemblages in burials at the cemetery. Thus, it is highly likely that burials containing snake skeletons, are a phenomenon *sui generis* for the early medieval period at Ravna–Slog and do not represent the Roman period and the Middle Ages separately.

As Danijel Džino suggests for Dalmatia in the early Middle Ages, the 'Old Croat' cemeteries are often located near or within Iron Age grave mounds or burial sites, ruins from Antiquity, in the vicinity of Late Antique churches or Late Antique cemeteries. In his interpretation, this is a clear sign of the local communities appropriating the local past for the construction of identity. He notes that these strategies indicate local communities and their elites had no control over the past of the land they had come to dominate, whose local elites lacking noble ancestry aimed to establish an ideological discourse or social system which would secure and justify social domination (Džino 2014, 139–140). The same sign of constructing identity by using the past may be seen in burials at Ravna–Slog through the placement of graves and reuse of earlier, late Roman material culture.

The early medieval reuse of Antique material culture is a familiar topic in European research (Williams 1997; Williams 1998, 92–96; Bradley 1998, 13–22; Gilchrist 2008, 139–144; Babić 2015, 248–262; Weiss-Krejci 2015, 307–309; Bradley 2015, 325–326; Curta 2016, 269–285). The culturally specific symbolic perception of ancient objects from Avar cultural contexts described by Aurel Rustoi suggests that such reused artifacts may have been handed down from generation to generation as heirlooms, playing a role in the remembrance of family tradition and ancestral identities, though perceived as elements of the natural environment (Rustoi 2015, 247–263).

The position of Ravna–Slog is complementary to numerous early medieval cases in terms of the reuse of "Antique" space or architecture for cemeteries in the Balkans (Grujić 2009; Veseličić 2009, 83–119; Ćirić 2014; Ćirić 2016, 731–747). Still, significant differences in the reuse of Roman material culture as well as the deliberate manipulation of animal bodies indicate that there was a strong pressure to form new local identities in this community. Locals needed to find "roots" during times of general social crisis in the Balkans during the Early Middle Ages. However, the reuse of objects and the peculiar treatment of wild animals, including snakes, in burial practices indicate that specific elements of the local culture were appropriated into those from the past (*sensu* Weiss-Krejci 2015, 307–309).

Conclusions

The positioning of snakes in the Ravna–Slog burials was the result of deliberate human action, an evidence of manipulating reptilian bodies in a period of conversion in the Christian world in early medieval Serbia. This conclusion is based on an examination of snake skeletons in graves compared against the reuse of material culture from the late Roman past.

It needs to be further investigated whether the graves containing snake skeletons date exclusively to the early Middle Ages and not to the Roman period. However, this study is a warning that zooarcheological finds and their taphonomic histories must always be considered when excavating and documenting burial sites. Due to the dearth of reported snake burials, it is possible that this curious zooarchaeological phenomenon has frequently been overlooked.

The Ravna–Slog case also helps reflecting on archeological tradition and field methods hitherto used. Archaeological excavations at Ravna–Slog were subject to the theoretical background of the cultural–historical approach, the snake remains having been attributed to natural causes. A vital clue thus may have been missed although this omission was not the fault of the researchers at the time.

Interpreting the practice of interring a woman or child individually with a snake imply aspects of dying, pain, and grief in this early medieval community. If the snake remains reflect key social relationships of the deceased after death, it is unclear how important relations between humans and animals were to the community and whether gender played a role in human–animal associations seen in the burial assemblages.

From a post-cultural history perspective, snakes may be seen as cultural deposits, factors in the burials of the local community. This phenomenon is compounded by the high number of infant graves. Children buried with snakes tend to be older than one year of age. This may imply that they were more socially integrated than younger infants.

In conjunction with the reuse of Roman artifacts and choice of burial locations in the Roman cemetery near the Roman and Early-Byzantine fortress of *Timacum Minus*, intense social pressures may be hypothesized compared to other contemporaneous communities. The human skeletons show poor living conditions, inadequate medical practices, and malnutrition.

The "reuse" indicators, manipulation of animal bodies and connotative practices at the wake of the Avar Khaganate may suggest dissatisfaction in a time of hardship, a desire to reflect on past traditions. Such ritual practices can be a response to change, even oppression, although, as is evident in the case of Ravna–Slog, it is difficult to study them archaeologically.

Current cultural memory research in archaeology generally deals with the commemoration of the dead, the most powerful aspect of social memory: they represent the past as remembered, constructing continuity in cultural memory. A retrospective remembrance of the dead is one way the group continues to live with their "ancestors" in the present. The call upon the past, as through the ancestors' reverence, is one means of constructing a cultural identity that also harmonizes with the present by reshaping the authenticity of the origins granted by the ancestor—itself a repurposing of the past. Social memory is often used to naturalize authority or in the service of resistance (Van Dyke and Alcock 2003, 3). When searching for cultural memories in archaeological contexts, the common norms of a community are to be identified by which the majority of society participated in a repetition of actions that can be recognized in the archaeological record.

If reusing material culture from the distant past is inherent to the reinforcement of cultural memory, it must be perceived as a communal practice. Furthermore, reuse may actually be part of the ritual. As opposed to otherwise extraordinary, rare, and exotic events, a ritual is what is repeated (Berggren and Nilsson Stutz 2010, 192). One snake put in one grave may be interpreted as exceptional, but a repetition of the practice is significant. According to Catherine Bell, ritualization comprises picking out a limited specification of actions and making them remarkable (Bell 1992; Gosden and Lock 1998, 4). The repetition of practices in early medieval burial rites once characteristic of the Avar Khaganate suggests more than just deliberate cultural action. Despite the questionable reliability of the data from older excavations, the practice of interring a woman or child individually with a snake is remarkable. With the snake deliberately straightened by their sides or wrapped around the waist of the deceased can be a ritualistic response of the Ravna–Slog community to social pressures; a resistance to change in a time of early medieval transition. These conclusions, even if limited, may help drafting a frame of reference for early medieval cemetery excavations in the future.

Acknowledgments This research was undertaken for project No 177008, funded by the Ministry of Education, Science and Technological Development of the Republic of Serbia. For key points, based on bioarchaeological methodology, appreciation is expressed to Ivana Živaljević from the Laboratory of Bioarchaeology in Belgrade as well as to my colleagues from the ZOOARCH mailing list (covering academic discussions of zooarcheology). Additional thanks go to herpetologist Rastko Ajtić and to the "Milutin Radovanović" Serbian Herpetological Society. I would like to express my gratitude to Sofija Petković for assistance with the unpublished excavation documentation. A final note of thanks goes to Danijel Džino for providing support for my writings on the reuse of material culture in the Middle Ages in Serbia.

References

Andersson Cederholm, Erika, Amelie Björck, Kristina Jennbert, and Ann-Sofie Lönngren eds. 2014. *Exploring the Animal Turn, Human-Animal Relations in Science, Society and Culture*. Lund: Lund University.

Albarella, Umberto, Mauro Rizzetto, Hannah Russ, Kim Vickers, and Sarah Viner-Daniels eds. 2017. *The Oxford handbook of Zooarchaeology*. Oxford: Oxford University Press.

Asman, Alaida. 2011. *Duga senka prošlosti. Kultura sećanja i politika povesti*. Beograd: XX vek.

Assmann, Jan. 2011. *Cultural Memory and Early Civilization: Writing, Remembrance, and Political Imagination*. Cambridge: Cambridge University Press.

Babić, Staša. "Biography of the Hill. Novi Pazar in South-Western Serbia." In *The Lives of Prehistoric Monuments in Iron Age, Roman and Medieval Europe*, edited by Marta Díaz-Guardamino, Leonardo García Sanjuán, and David Wheatley, 249–264. Oxford: Oxford University Press, 2015.

Bartosiewicz, László, Zsófia Eszter Kovács, and Balázs Farkas. "Pass the Skeleton Key...Animals in an Early Copper Age Inhumation Burial from Pusztataskony–Ledence I, Hungary." In *Uncomformist Archaeology: papers in Honour of Paolo Biagi*, edited by Elisabeta Starnini, 77–88. British Archaeological Reports, International Series 2528. Oxford: Archaeopress, 2013.

Bell, Catherine. 1992. *Ritual Theory, Ritual Practice*. Oxford: Oxford University Press.

Berggren, Åsa, and Liv Nilsson Stutz. "From spectator to critic and participant. A new role for archaeology in ritual studies." *Journal for Social Archaeology* 10, no. 2 (2010). 171–197.

Bradley, Richard. "Ruined buildings, ruined stones: Enclosures, tombs and natural places in the Neolithic of south-west England." *World Archaeology* 30, no. 1 (1998). 13–22.

Bradley, Richard. "Piecing together a Past." In *The Lives of Prehistoric Monuments in Iron Age, Roman and Medieval Europe*, edited by Marta Díaz-Guardamino, Leonardo García Sanjuán, and David Wheatley, 325–342. Oxford: Oxford University Press, 2015.

Chapman, Robert, and Alison Wylie. 2016. *Evidential Reasoning in Archaeology*. London: Bloomsbury.

Ćirić, Gordana. (2014). "*Roman coins out of time: the transformation of values. The reuse of Roman coins in medieval cemeteries (AD 400–1400) in the territory of Serbia.*" PhD diss., Johan Wolfgang Goethe-Universität zur Frankfurt, 2014.

Ćirić, Gordana. "Rimski novčići u ženskim srednjovekovnim grobovima sa teritorije Srbije: mogućnosti interpretacije." *Etnoantropološki problemi* 11, no. 3 (2016). 731–747.

Curta, Florin. "Burials in Prehistoric Mounds: Reconnecting with the Past in Early Medieval Greece." *Revue des Études Byzantines* 74 (2016). 269–285.

Curta, Florin. "Church, churchyard, and children in early Medieval Balkans: a comparative perspective." In *Migration, Integration and Connectivity on the Southeastern Frontier of the Carolingian Empire, East Central and Eastern Europe in the Middle Ages, 450–1450*, edited by Danijel Dzino, Ante Milošević, and Trpimir Vedriš , 245–267. Leiden: Brill, 2018.

Đorđević, Sonja, and Ljiljana Tomović. "Dolichophus caspius." In *Red Book of Fauna of Serbia II – Reptiles*, edited by Ljiljana Tomović, Miloš Kalezić, and Georg Džukić, 213–219. Belgrade: Institute for Nature Conservation of Serbia, 2015.

Đorđević, Tihomir R. 1958. *Priroda u verovanju i predanju našeg naroda II. Srpski etnografski zbornik, Knjiga LXXII*. Beograd: Srpska akademija nauka.

Džino, Danijel. 2010. *Becoming Slav, Becoming Croat. Identity Transformations in Post-Roman and early Medieval Dalmatia*. Leiden: Brill.

Džino, Danijel. "The Rise and Fall of Dalmatian 'Big man': social structure in Late Antique, Post-Roman and Early Medieval Dalmatia (ca. 500–850)." In *Studia Academica Šumensia 1, The Empire and Barbarians in South-Eastern European in Late Antiquity and Early Middle Ages*, edited by Stoyan Vitlyanov, and Ivo Topalilov, 127–152. Shumen: The University of Shumen Press, 2014.

Effros, Bonnie. 2003. *Merovingian Mortuary Archaeology and the Making of the Early Middle Ages*. Berkeley: University of California Press.

Ercegović-Pavlović, Slavenka. 1980. *Les Nécropoles Romaines et Médiévales de Mačvanska Mitrovica*. Sirmium XII. Beograd: Arheološki institut.

Ferjančić, Božidar. 2009. *Vizantija i Južni Sloveni*. Beograd: Ethos.

Fleck, Ludwig. "To look, to see, to know." In *Condition and Facts Materials on Ludwik Fleck, Boston studies in the Philosophy of Science*, edited by Robert S. Cohen, and Thomas Schnelle, 129–152. Boston: D. Reidel, 1986 (1947).

Gilchrist, Roberta. "Magic for the Dead? The Archaeology of Magic in Later Medieval Burials." *Medieval Archaeology* 52 (2008). 119–159.

Gosden, Chris, and Gary Lock. "Prehistoric histories." *World Archaeology* 30, no. 1 (1998). 2–12.

Grujić, Ivan. 2009. *Srednjovekovne nekropole na prostoru antičkih sakralnih objekata: master rad*. Beograd: Filozofski fakultet, Univerzitet u Beogradu.

Härke, Heinrich. "Grave goods in early medieval burials: messages and minings." *Mortality* 19, no. 1 (2014). 41–60.

Hill, Erica. "Archaeology and Animal Persons, Toward a Prehistory of Human-Animal Relations." *Environment and Society: Advances in Research* 4 (2013). 117–136.

Ilijić, Bojana. 2015. *Timacum Minus – dva sveta*. Knjaževac: Zavičajni muzej Knjaževac.

Kiss, Gábor. "A závodi avarkori temető." In *Tolna megyei avar temetők. Wosinsky Mór ásatásai*, edited by Gábor Kiss, and Péter Somogyi, 161–185. Dissertationes Pannonicae III, no. 2. Budapest: Eötvös Loránd Tudományegyetem, 1984.

Kopytoff, Igor. "The cultural biography of things: commodization as process." In *The social life of things. Commodities in cultural perspective*, edited by Arjun Appadurai, 64–94. Cambridge: Cambridge University Press, 1986.

Lenders, H. J. Rob, and Ingo A. W. Janssen. "The Grass Snake and the Basilisk: From Pre-Christian Protective House God to the Antichrist." *Environment and History* 20 (2014). 319–346.

Marjanović-Vujović, Gordana. "Rural settlements in the 9th and 10th centuries in the Danube Valley in Serbia." In *From the Baltic to the Black Sea, Studies in medieval archaeology*, edited by David Austin, and Leslie Alcock, 236–247. London: Routledge, 1997.

Mikić-Antonić, Branislava. 2012. *Nekropola iz perioda avarske dominacije*. Bečej: Gradski muzej Bečej.

Milosavljević, Monika. "Ludwik Fleck's concepts slicing through the Gordian Knot of Serbian Archaeology." *Transversal, International Journal for Historiography of Science* 1 (2016). 88–100.

Ostrogorsky, Georgy A. 1968. *History of the Byzantine State*. Oxford: Basil Blackwell.

Petković, Sofija, Mira Ružić, Svetozar Jovanović, Marko Vuksan, and K. Zsuzsanna Zoffmann. 2005. *Roman and Medieval Necropolis in Ravna near Knjaževac*. Beograd: Arheološki institut.

Petković, Sofija, and Nataša Miladinović-Radmilović. "Neki primeri pogrebnog kulta na kasnoantičkoj nekropoli u Ravni (Timacum Minus) kod Knjaževca." *Etno-kulturološki zbornik za proučavanje istočne Srbije i susednih oblasti* XVII (2013). 47–60.

Petković, Sofija, Bojana Ilijić, and Nikola Radinović. "Nekropola Slog – Timacum Minus u Ravni kod Knjaževca." In *Zaštitna arheološka iskopavanja 2013. godine. Arheologija u Srbiji, projekti Arheološkog instituta u 2013. godini*, edited by Dragana Antonović, 76–81. Beograd: Arheološki institut, 2014.

Petković, Sofija, Jelena Bulatović, and Dragica Bizjak. "Jedna zanimljiva sahrana sa nekropole Slog u Ravni." In *XXXVIII skupština i godišnji skup Srpskog arheološkog društva (Pirot, 4–6. jun 2015. godine), Knjiga apstrakata*, edited by Vojislav Filipović, and Dragana Antonović, 89–90. Pirot: Srpsko arheološko društvo, 2015a.

Petković, Sofija, Nadežda Gavrilović-Vitas, Nataša Miladinović-Radmilović, and Bojana Ilijić. 2016. *Pogrebni ritual i Dionisov kult u Ravni (Timacum Minus)*. Beograd: Arheološki institut.

Petković, Sofija, Dragica Bizjak, and Jelena Bulatović. "Early Medieval Burial of Woman and Fox at the Necropolis Slog in Ravna (Timacum Minus) in Eastern Serbia". In "*Homines, Funera, Astra*", 2015b. Alba Iulia: Universitatea "1 Decembrie 1918" (*in press*).

Pluskowski, Aleks ed. 2007. *Breaking and Shaping Beastly Bodies: Animals as Material Culture in the Middle Ages*. Oxford: Oxbow Books.

Pohl, Walter. "A Non-Roman Empire in Central Europe: The Avars." In *Regna and Gentes, The Relationship between Late Antique and Early Medieval Peoples and Kingdoms in the Transformation of the Roman World*, edited by Hans-Werner Goetz, Jörg Jarnut, and Walter Pohl, 571–595. Leiden: Brill, 2003.

Potts, Daniel T. "Revisiting the snake burials of the Late Dilmun building complex on Bahrain." *Arabian Archaeology and Epigraphy* 18, no. 1 (2007). 55–74.

Premužić, Zrinka, Petra Rajić Šikanjić, and Anita Rapan Papeša. "Eggshells in a juvenile burial from Avar cemetery in Croatia." In *Abstract book of the 19th Annual Meeting of the European Association of Archaeologists*, edited by Jan Turek, 58–59. Pilsen: University of West Bohemia in Pilsen, 2013.

Reitz, Elizabeth J., and Elizabeth S. Wing. 2008. *Zooarchaeology (Second edition)*. Cambridge: Cambridge University Press.

Ritvo, Harriet. "On the animal turn." *Daedalus* 136, no. 4 (2007). 121.

Rosner, Gyula. "Előzetes jelentés a Gyönk–Vásártér úti avar temető feltárásáról II." *A Szekszárdi Balogh Ádám Múzeum Évkönyvéből 1971–1972/Különnyomat* no. II–III (1972). 85–221.

Rustoiu, Aurel. "Magic and Memory. Prehistoric and Roman 'Antiquities' in Avar Period Graves from the Carpathian Basin." *Transylvanian Review* XXIV, no. 2 (2015). 247–263.

Špehar, Perica N. 2017. *Centralni Balkan od 7. do 11. veka, arheološka svedočanstva*. Beograd: Booka.

Szentpéteri, József. "Kígyómotívum a griffes-indás népesség hagyatékában." *A Herman Ottó Múzeum Évkönyve* 30–31, no. 2 (1993). 225–278.

Tomović, Ljiljana, Rastko Ajtić, Katarina Ljubisavljević, Aleksandar Urošević, Danko Jović, Imre Krizmanić, Nenad Labus, Sonja Đorđević, Miloš L. Kalezić, Tanja Vukov, and Georg Džukić. "Reptiles in Serbia: Distribution and diversity patterns." Bulletin of the Natural History Museum 7 (2014). 129–158.

Tomović, Ljiljana and Georg Džukić. "Elaphe quatuolineata." In *Red Book of Fauna of Serbia II – Reptiles*, edited by Liljana Tomović, Miloš Kalezić, and Georg Džukić, 220–226. Belgrade: Institute for Nature Conservation of Serbia, 2015.

Tomović, Ljiljana, and Dimitar Lakušić. "Habitat of Reptiles in Serbia.", In *Red Book of Fauna of Serbia II – Reptiles*, edited by Liljana Tomović, Miloš Kalezić, and Georg Džukić, 49–63. Belgrade: Institute for Nature Conservation of Serbia, 2015.

Todd, Josie, Joshua Amiel, and Richard J. Wassersug. "Factors influencing the emergence of northern populations of Eastern Ribbon Snakes (*Thamnophis sauritus*) from artificial hibernacula." *Canadian Journal of Zoology* 87, no. 12 (2009). 1221–1226.

Van Dyke, Ruth M. and Susan E. Alcock. "Archaeologies of Memory: An Introduction." In *Archaeology of Memory*, edited by Ruth M. Van Dyke, and Susan E. Alcock, 1–13. Melbourne: Blackwell, 2003.

Veseličić, Milica. "Sahranjivanje tokom srednjeg veka i novijeg doba u praistorijskim humkama u čačanskom kraju." *Zbornik radova Narodnog muzeja u Čačku* XXXVIII (2009). 83–120.

Weiss-Krejci, Estella. "The Plot Against Past. Reuse and Modification of Ancient Mortuary Monuments as Persuasive Efforts of Appropriation." In *The Lives of Prehistoric Monuments in Iron Age, Roman and Medieval Europe*, edited by Marta Díaz-Guardamino, Leonardo García Sanjuán, and David Wheatley, 307–324. Oxford: Oxford University Press, 2015.

Williams, Howard. "Ancient Landscapes and the Dead: The Reuse of Prehistoric and Roman Monuments as Early Anglo-Saxon Burial Sites." *Medieval Archaeology* 41 (1997). 1–32.

Williams, Howard. "Monuments and the past in early Anglo-Saxon England." *World Archaeology* 30, no. 1 (1998). 90–108.

Živaljević, Ivana. "Životinje između Prirode i Kulture: priča o arheozoologiji." *Etnoantropološki problemi* 8, no. 4 (2013). 1137–1163.

CHAPTER 7

Perpetual Preys: Pursuing the Bonnacon Across Space and Time

Zsuzsanna Papp Reed

Introduction

Projectile defense mechanisms of non-human animals have intrigued their human counterparts for centuries if not more—and continue to do so. J. K. Rowling's blast-ended skrewts and the perennially popular internet lists of "The Most Disgusting Animal Defenses," normally containing sensationalist descriptions of the fulmar, the hoopoe, and the pygmy sperm whale, look back to a long history. Humans, proud of their own prowess in inventing and yielding projectile weapons, have recorded the natural occurrence of this ability in animals in various ways and for various purposes. This brief essay addresses medieval Latin evidence for one such animal—or three, even four, depending on how one looks at it.

As far as we know, the beast—let us call it the *bonasus* for now—was first described by Aristotle in the fourth century BC in a detailed account of the appearance and character of an animal called *bonasos* (βόνασος, also appears as βόνασσος and βώνασος) by the Greeks and *monapos* (μόναπος)

Z. Papp Reed (✉)
Department of Medieval Studies, Central European University, Vienna, Austria
e-mail: reedzs@ceu.edu

L. Bartosiewicz, A. M. Choyke (eds.), *Medieval Animals on the Move*, https://doi.org/10.1007/978-3-030-63888-7_7

by the Paeonians to whose lands it was native (roughly present-day North Macedonia). The animal is certainly odd: it is large and has a mane of a horse, its "horns are crooked, curved towards each other and not useful for defence," and its skin is thick, its meat pleasant, so it is often hunted. Most importantly: "it defends itself by kicking and by voiding dung at them, throwing it up to four rods from itself," which it uses frequently and with ease and which "scorches the hounds' coats so that they rub off" (Book 8 (9), ch. 45 in Aristotle 1991, 3: 386–391). Thus, the *bonasus* is a large undomesticated bovine which defends itself using excrement that is either hot or acidic. It needs to resort to this frequently as it is hunted for its meat, and possibly for its giant hide and horns as well. Aristotle's description is vivid and detailed, and it obviously sounds like an actual animal: even laymen would conclude that it is about some sort of a large bison or buffalo. Due to the meticulous portrayal, and probably to the scatological detail, the animal's description remained intriguing and challenging for naturalists, philologists, archaeologists, and historians alike.

The Pursuit

Understanding what exactly the *bonasus* was (or is) is a complicated issue. First, what is available is a description and not the animal itself. Having neither Aristotle nor the *bonasus* at hand, the reader sees through a mirror, darkly. To make up for the absence of both Aristotle and the beast, one can turn to similar descriptions, images, and bones. There is a wealth of often rather similar (more often conflicting) animal descriptions found in classical and medieval sources, and all the known images are associated with these. From the sixteenth century onward, generations of naturalists applied themselves to identify the beast and create order in the chaos of *bonasus*, *bubalus*, *bisontes*, *bos indicus*, *zubr*, even the Aeolian bull whose horns could seat three grown men, based on the scanty and often conflicting ancient descriptions of size, color, build, and distinguishing features (Gesner et al. 1551; Wotton 1552; Buffon and Barr 1807, 8, 19–58).

While identifying the animals in the text with real animals is no doubt a rewarding exercise and certainly contributed to our knowledge of the aurochs for example, it never quite got anywhere when it came to the *bonasus*. After the fruitless attempts to certifiably "identify" the animal in the last hundred years, modern zoology and archaeozoology stepped away from the old yarns about the *bonasus* entirely. In scholarship, it came to be defined as a hybrid of medieval bestiary lore. Along the likes of the hippogriff or the monoceros, it is characterized as one of the fantastic hybrids of bestiaries, that of a horse and a bull (Clark 2006, 36; Werness 2007, 34).

The *bonasus*, thus, somehow fell through the cracks. Before concluding that it was a mythical hybrid, scholarship had been largely preoccupied with identifying and describing "real" animals by drawing together disparate descriptions of animals appearing under various names and often described very differently—in ancient texts, early modern zoological observations, and modern archaeology. Instead of giving up on the *bonasus* as a figment of imagination, our question here is what Susan Crane formulated about the main preoccupation of medieval naturalists and bestiaries which "focus on the perceptual process": "what has been *thought* about animals? What do they *mean* to us? In consequence, the bestiary's animals have mixed ontological status: they may or may not exist in the world, but they certainly exist in cultural memory" (Crane 2013, 88). Rather than attempting to gather multiple pieces of textual evidence to arrive at one animal, in this study, the description of "the *bonasus*" becomes a multiple entity through the editorial and authorial practices, even the errors and deletions, of the medieval and early modern scribes and scholars describing it.

The medieval tradition goes back to Aristotle—with a little help from Latin Antiquity. The *bonasus/bonacus*, obviously the same animal in these descriptions, begins its long medieval journey in the descriptions of first-century Pliny and third-century Solinus, who both echo Aristotle's description, albeit in Latin and in a somewhat abbreviated form. Pliny retains the information that the *bonasus* lives in Paeonia, but Solinus 200 years later moves its habitat to Phrygia, the west central part of Anatolia in present-day Turkey. In Table 7.1, the text is broken up to sentences and phrases to emphasize the commonalities and differences, but the order is not altered. The information is presented in precisely the same order in all three texts, which is a commonality itself.

As Buffon noted in the second half of the eighteenth century, "what must appear singular, the *bonasus*, although fully described by the great philosopher [Aristotle], has not been recognised by any of the Greek and Latin naturalists who have written after him, all of whom have literally copied him on this subject; so that to this day, there is no more than the name of the *bonasus* known, without the knowledge of the animal which it ought to be applied to." (Buffon and Barr 1807, 8: 21). The hiatus is indeed interesting: the Early Middle Ages gave rise to tremendously popular and widely disseminated texts in which one would expect to find the *bonasus*, for example, the third- or fourth-century Greek *Physiologus* or Isidore's *Etymologies*, but none of them contained information about this

Table 7.1 Commonalities and differences between *bonasus* descriptions in three sources

Aristotle *(Translation in Aristotle 1991, 3:386–90)*	*Pliny, Naturalis historia, 8.16.* *(Translation in Pliny the Elder 1855)*	*Solinus, Collectanea rerum memorabilium, 40:10 (Solinus 1895, 168), my translation.*
The bison [βόνασος] occurs in Paeonia on the Messapian mountain which marks the boundary between the Paeonian and the Maedican country. The Paeonians call the animal *monapos* [μόναπος].	[...] *tradunt in Paeonia feram quae bonasus vocetur.* In Pæonia, it is said, there is a wild animal known as the *bonasus.*	*Phrygia. [...] in his locis animal nascitur quod bonacum dicunt,* Phrygia. [...] In this place an animal is found which is called the *bonacus,*
Its size is that of a bull; it is more bulky than the ox, for it is not elongated. Its skin when stretched out covers a seven-seat dining-room. Its form is generally like an ox except that it has a mane down to the shoulder like a horse; but the hair is softer than the horse's and lies closer. The color of the hair is fair and the mane is deep, reaching down to the eyes, and thick. But the animal's color is a mean between ashen and red, unlike the horses called *paroan*; its hair however is rather rough and down below it is woolly. They do not occur as black or very red.	*equina iuba, cetera tauro similem,* it has the mane of the horse, but is, in other respects, like the bull.	*cui taurinum caput ac deinceps corpus omne, tantum iuba equine.* and whose head, and thereafter whole body are of the size of a bull's with the mane of a horse.

(*continued*)

Table 7.1 (continued)

Aristotle (Translation in Aristotle 1991, 3:386–90)	*Pliny, Naturalis historia, 8.16. (Translation in Pliny the Elder 1855)*	*Solinus, Collectanea rerum memorabilium, 40:10 (Solinus 1895, 168), my translation.*
Their voice is like an ox; their horns are crooked, curved toward each other and not useful for defense, in size a span or a little more, in thickness such as to hold not much less than a half-pitcher each; the horn's dark color is beautiful and shiny. The forelock reaches down to the eyes, so that the animal moves off to one side rather than to the front. It lacks the upper teeth, like the ox and all horned animals; its legs are shaggy; it is cloven-hoofed; its tail is too small in proportion to its size, and resembles that of the ox. It throws up the dust and digs like a bull. Its skin is strong in resisting blows. Its meat is pleasant, and therefore they hunt it.	*cornibus ita in se flexis, ut non sint utilia pugnae.* with horns, however, so much bent inward upon each other, as to be of no use for the purposes of combat.	*cornua autem ita multiplici flexu in se recurrentia, ut si quis in ea offendit, non vulneretur.* its horns are twisted manifold, curling back on themselves in such a way that if anyone attacks it, it is not harmed.
On being hit, it runs away; it stands to meet them when it has lost strength.	*quapropter fuga sibi auxiliari* it has therefore to depend upon its flight.	*sed quicquid praesidii monstro illi frons negat, alvus sufficit.* but the protection which its forehead denies this monster is compensated by its bowels.

(*continued*)

Table 7.1 (continued)

Aristotle (Translation in Aristotle 1991, 3:386–90)	*Pliny, Naturalis historia, 8.16. (Translation in Pliny the Elder 1855)*	*Solinus, Collectanea rerum memorabilium, 40:10 (Solinus 1895, 168), my translation.*
It defends itself by kicking and by voiding dung at them, throwing it up to four rods from itself.	*reddentem in ea fimum, interdum et trium iugerum longitudine* and, while in the act of flying, it sends forth its excrements, sometimes to a distance of even three *jugera.*	*nam cum in fugam vertit, proluvie citi ventris fimum egerit per longitudinem trium iugerum,* For when it turns to flee, it quickly discharges dung from its stomach over a distance of three *iugera.*
It uses this easily and frequently, and it scorches the hounds' coats so that they rub off. Now it is when the bison is distressed and frightened that its dung does this, but when it is free from distress, it does not scorch. Such then is the beast's form and nature.	*cuius contactus sequentes ut ignis aliquis amburat* the contact of which burns those who pursue the animal, just like a kind of fire.	*cuius ardor quicquid attigerit amburit* the heat of which burns anything it touches.
When it is time to give birth, they give birth in a group in the mountains. They first drop dung around the place before giving birth, and make a sort of rampart; for the beast discharges a great quantity of this residue.		*ita egerie noxia submovet insequentes.* in this way, it keeps away pursuers with its harmful excrement.

creature. The Plynian/Solinian passage was discontinued for a couple of centuries until it was rediscovered by the creators of the medieval bestiary tradition—still repeating the same text.

There is no room or, indeed, need to survey the textual relationship and dissemination of bestiaries and bestiary-type texts in England and Europe here, as many recent works provide excellent overviews of the various families and the networks that can be reconstructed based on intertextuality and comparative analysis of the illustrations (McCulloch 1962; Baxter 1998; Clark 2006; Dines 2012; Stewart 2012). The earliest Latin versions of the *Physiologus* from the fifth century were the predecessors of medieval bestiaries. Corresponding to the distinct lack of the *bonasus* in early medieval sources, the so-called B-Is type bestiaries—essentially the Latin *Physiologus* expanded by passages from Isidore's *Etymologies*—obviously do not contain the description of the *bonasus*. The ones that do contain the animal now called a *bonnacon* (or *bonacon*) are the Transitional Bestiaries (Dines 2012), as well as the Second and some Third Family bestiaries, which all derive from the B version of the *Physiologus* but are expanded by copious additions from other sources.

The Second Family was the only bestiary version that initially developed in England before travelling to the continent (Clark 2006, 12, 24). Much longer than the original *Physiologus*, these bestiaries contain over a hundred animals across nine major divisions of varying length. Not all Second Family bestiaries contain the *bonnacon/bonacon* entry, for example, Stowe 1067 does not, while its cognate Morgan MS does (Baxter 1998, 106–107). In these texts, the *Physiologus* core was furnished with extra material in the 1160s or 1170s, including the entry about the *bonnacon* following Solinus's description nearly word by word (Clark 2006, 21–22). Even though its description remains the same, the bestiary tradition expands the habitat of the *bonnacon*. The beast is now located in Asia instead of the narrower geography of Paeonia (Aristotle and Pliny) and Phrygia (Solinus). Also note the slightly different name, consistent across all manuscripts:

> In Asia an animal is found which men call *bonnacon*. It has the head of a bull, and thereafter its whole body is of the size of a bull's with the maned neck of a horse. Its horns are convoluted, curling back on themselves in such a way that if anyone comes up against it, he is not harmed. But the protection which its forehead denies this monster is furnished by its bowels. For when it turns to flee, it discharges fumes from the excrement of its belly over a distance of three acres, the heat of which sets fire to anything it touches. In this way, it drives off its pursuers with its harmful excrement. (The Aberdeen Bestiary, Aberdeen University Library MS 24, fol. 12r)

Later, the so-called Third Family bestiaries show another recasting of the *Physiologus*. Ilya Dines argues that the main source of the Third Family moralizations was the *Distinctiones* by the Lincoln chancellor, William de Montibus (Dines 2008). These, surviving in only five thirteenth-century manuscripts, contain much of the same material as the Second Family books, but in a different order and with entirely different supplementary texts: in Cambridge, Fitzwilliam MS 254 (ca 1220–1230), the *bonnacon* is on fols 21rv; London, Westminster Abbey MS 22, fols 25rv; (ca 1275–1300, from the Franciscan Friary in York); in Oxford, Bodleian Library MS Douce 88, fols. 68–154, the *bonnacon* is on fol. 92v (NB: fols 5r-29r in this volume are a Second Family bestiary, in which the *bonnacon* is on fol. 10r); and Oxford, Bodleian Library MS e Mus. 136 (ca 1260–80, possibly Netherlandish), in which the *bonnacon* is on fol. 24v. The manuscript in Cambridge University Library MS Kk.4.25 (ca 1220–40) does not contain an entry on the *bonnacon* (McCulloch 1962, 38–39; Baxter 1998, 132–33). The *bonnacon* is normally placed after the hyena but not always in the same place in different manuscripts (Baxter 1998, 140). Four manuscripts contain a longish, moralized account, based on the Solinian text. Citing Jeremiah 3:3, "And thou had a whore's forehead; thou refused to be ashamed," the moralization likens the animal to an impudent person, which, according to Dines, probably comes from a distinctly non-bestiary context: either Gregory the Great's *Homiliae in Hiezechielem*, or Hrabanus Maurus's commentary (Dines 2008, n. 898). All Third Family illustrations show a hunter protecting himself with a shield, thrusting a spear in the animal's side.

The Fourth Family bestiary, associated with the bestiary chapter in the *De proprietatibus rerum* by Bartholomaeus Anglicus, is represented by one surviving manuscript in the Cambridge University Library: MS Gg. 6.5 (fifteenth century). The bestiary is on folios 1r-100r and, as Florence McCulloch (1962, 39) observes, "the large, careless pictures in this manuscript look like poor relations of those found in Copenhagen, Kongelige Bibliotek Gl. Kgl. 1633 4° (a Second Family manuscript; Fig. 7.1)." The text of the bestiary is based on Book 18, "*De animalibus*," of the *De proprietatibus rerum*, so it does not contain the description of the *bonasus/ bonnacon*. As will be discussed below, Bartholomaeus Anglicus did include a variant description of the *bonnacon* elsewhere in his encyclopedia, this, however, was not part of the text transmitted into the Fourth Family bestiary.

Before moving on to the next stage (and next geographical region) in the beast's medieval career, it is worth briefly mentioning that the Solinian version also appears near Phrygia on the Hereford Map (ca 1300) with the caption "In Phrygia there is born an animal called *bonnacon*; it has a bull's head, horse's mane and curling horns, when chased it discharges dung over an extent of three acres which burns whatever it touches" (Westrem 2001, 111). This information either preceded or disregarded the bestiary tradition, which was in full swing by this time: the beast being located in

Fig. 7.1 *Bonacon* in a Second Family Bestiary, fifteenth-century English manuscript (Copenhagen, Kongelige Bibliotek, Gl. kgl. S. 1633 4°, Folio 10r. Image in the public domain). http://www.kb.dk/permalink/2006/manus/221/eng/10+recto/?var

Phrygia echoes Solinus rather than the more vague geography indicated in the bestiaries (Jones 2013, 58). It also appears on the famous Ebstorf Mappamundi in the form of an animal attacking its pursuers, bearing the one-word rubric "bonasus," near Pamphylia, the region directly west of Phrygia (EbsKart at Leuphana Universität Lüneburg n.d.). Sidestepping the bestiaries' generic "Asia" as the native land of the animal, the location of the simple figure on the Ebstorf map goes back to either the Phrygian *bonnacon* of the Hereford map, or directly to Solinus.

An interesting pair of animals, the *elles* and the *urus* (elk and aurochs), inhabiting *Ruscia regio* on the Ebstorf map turns the hunting expedition for our beast toward the frosty northeast. While twelfth- and thirteenth-century bestiaries persistently featured the dung-spraying Asian *bonnacon*—sometimes with illustrations showing the animal in action—various geographical, historical, and scientific accounts added to the range of interesting animals living on the margins of the Christian *oikumene*, for example, the elk and the aurochs. At around this time, the concept of the wild ox was enriched with new materials which entered knowledge production through the rediscovery of Aristotle, access to translated works of

Ibn Sina, and the birth of the encyclopedia as we know it. The encyclopedic material about the fauna of the world was clearly expanding beyond the traditional framework of the *Physiologus* and bestiaries: the Aristotelian *bonasos* remained, and new, surprisingly similar, animals appeared alongside it.

Four influential encyclopedists are of note at this stage. They were born less than 50 years apart and moved around in cultural-intellectual circles that often more than tangentially overlapped. The outlier is the English Gervase of Tilbury (ca 1154–1227), whose career as a courtier of Holy Roman Emperor Otto IV took him from Essex to Sicily, Bologna, Rheims, Arles, and Braunschweig. He was born earlier than the others, a secular author whose encyclopedia was not really an encyclopedia as we know it, but one (to borrow Van Oppenraay's words about the similar *Hortus sanitatis*), that was designed for the lay aristocracy's entertainment with the *mirabilia* and *curiositates naturae*, and stories about magic rituals with animals (van Oppenraay 2015, 4). The other three were of the same age and all mendicants: the German Albertus Magnus (ca 1200–1280) who spent his life studying and teaching at universities and Dominican convents in Padua, Bologna, Cologne, Regensburg, Freiburg, Strasbourg, Hildesheim and Paris; Albertus's disciple, the Brabantian Thomas of Cantimpré (1201–1272), who entered the Dominican order at Leuven, but also travelled to Cologne, and was engaged in missionary activities throughout Brabant, Germany, Belgium, and France; and the English-born Franciscan Bartholomaeus Anglicus (1203–1272), a student at Oxford and lecturer at Paris, who travelled extensively in Saxony, Austria, Bohemia, and possibly Poland. It is perhaps also notable here that Michael Scot, translator of Aristotle from Arabic, worked in the Hohenstaufen court as Emperor Frederick II's court physician, and astrologer, and is known to have been in Bologna in 1220 (van Oppenraay 2015, 4–5).

The early thirteenth-century *Otia imperialia* (Recreation for an Emperor) by Gervase of Tilbury was written for Prince Henry, son of Henry II of England. Among the wealth of material regarding what would be now called life sciences, folklore, and ethnography, the book includes bestiary type descriptions inserted into the section entitled "The Resourcefulness of Animals." In one of these descriptions, Gervase describes onagers as indigenous to Poland. The onagers of Antiquity (E.g. Varro, *De re rustica*, 2.6.2-3) live in Phrygia and their attributes in the bestiary tradition are that the male castrates its foals out of jealousy, and that they mark the equinox by braying twelve times. While Gervase's

Polish onager does not correspond to late Antique and Isidorean accounts at all, it certainly bears vague resemblance to the Solinian *bonacus*, and the *bonnacon* of medieval bestiaries:

> 3.65 [...] *Porro apud Polloniam onagrorum frequens est uenatio, quorum eadem est fere natura que et asinorum: sunt enim asini siuestres cornua habentes ad modum damarum, melancolice complexionis; dum siccitatem habent innatam, sitibunda sunt. Cum ergo uenatorum cornua intonantia presentiunt, ad fontes properant, atque immenso haustu se saciant, ut siccitatemm ex calore currentis adauctam, sumpti potus humiditate temperent, Enimuero canibus urgentibus, paulatim gradum sistunt; et superuenientes tanta inundatione per nares inuoluunt quod et execatos diucius a sequendo retardant, et inter oculos obductos ad fugam citatam se accigunt. De hiis ad litteram in psalmo dicitur: 'Expectabunt onagri in siti sua.'*
>
> Further, in Poland onagers are often hunted. These animals are virtually identical to asses in their nature; they are in fact wild asses, with antlers like deer, and of a melancholic constitution. Since they have an innate dryness, they are thirsty creatures. Consequently, when they hear the warning of hunters' horns being sounded, they rush to the springs and fill themselves with an enormous draught of water, so that when their dryness is increased by the heat of running, they may counterbalance it with the wetness of the drink they have taken. Then, with the dogs pressing hard upon them, they gradually slacken their pace; when the dogs overtake them, they overwhelm them with such a great deluge from their nostrils that they both hinder them from pursuing them for some time by blinding them and give themselves the chance to run away quickly while the dogs' eyes are darkened. A psalm in its literal meaning says of these: 'The wild asses shall expect in their thirst.' (Gervase of Tilbury 2002, 684–85)

The Polish onager is, thus, characterized by its unique defensive projectile when hunted. Although the most important characteristics of the beast—habitat, name, and precise defense mechanism—are all markedly different from those of the *bonasus/bonnacon*, they are in fact loosely linked by way of another unusual description. Sometime later, probably in Paris, *De proprietatibus rerum* (On the properties of things) was written by Bartholomaeus Anglicus. In the alphabetized book on world geography (15), Bartholomaeus gives a general description of the lush and abundant lands of Bohemia, followed by a detailed list of native beasts, including a lengthier account that looks like a passage out of a bestiary: a beast that sprays its pursuers when distressed.

15.30 [...] *in eius montibus abundant pini & abietes, abundant & herbae innumerabiles non solummodo pascuales, verum etiam aromaticae & medicinales, seil, ursi, apri, cerui, capreole, tragelaphi, Bubali seu bisontes & inter has feras est quaedam habens magnitudinem bouis, haec bestia ferox est, saeua & habet magna cornua & ampla, cum quibus se defendit. Sed habens sub mento amplum folliculum, in ipso aquas recolligit, & currendo, aqua miro modo in illo folliculo calefacit, quam super venatores seu canes sibi nimis appropinquantes proiicit, & quicquid tetigerit depilat horribiliter & exurit.*

In its [i.e. Bohemia's] mountains pine and fir trees abound. Countless plants also abound, not only for grazing, but also spices and medicinal herbs. *Seil* [unidentified animal], bears, boars, deer, roes, wild goats, gazelles or bison abound, and among these beasts is a certain [beast] which is the size of an ox. This beast is fierce, wild, and has a long and wide horn, with which it defends himself. But it has under its chin a large bladder in which it collects water and when it runs, it warms the water in its bladder in the most marvellous way, which it then shoots at hunters or dogs coming too close to it. And this water horribly removes hair from and scalds whatever it touches. (Bartholomaeus Anglicus 1964, 639)

Although this animal sounds much like Gervase's Polish onager, and is certainly described in the same geographical area, Bartholomaeus abandons both the ass-like character and the misleading moniker, keeping the name onager for braying wild asses in his bestiary written in the veritable classical-Isidorean tradition (Bartholomaeus Anglicus 1964, 1096, 18.76). This Bohemian animal seems like an amalgamation of the bestiaries' Solinian *bonacus*, Gervase's onager, and Cantimpré/Albertus's *zubr* (discussed below): it is "the size of an ox" and "horribly removes hair from and scalds whatever it touches," which echoes the *bonasos*, which scorches the hounds' coats so that they rub off (Aristotle 1991, 3:391). At the same time, it sprays the hunters and their dogs with water from its front end just like the onager. Another, less direct commonality is heat being mentioned in relation to water: the onager drinks an enormous draught of water to make up for the heat it develops when running from its pursuers, the Bohemian beast fills up with an equally large amount of water, which it then heats up as he runs. The feature that is different in all three animals is their horns/antlers: the Solinian *bonacus* has horns with which it is unable to defend itself, Bartholomaeus's beast has "a long and wide horn, with which it defends himself" besides using the deadly *folliculum* under its chin, and Gervase's onager has antlers like a deer, much like the probably unrelated illustration in the Koninklijke Bibliotheek (KB, KA 16, Folio 47v. Fig. 7.2).

Bartholomaeus Anglicus's encyclopedia enjoyed such an enormous and long-lasting influence on subsequent writing, scientific and otherwise, that there is no space to present the afterlife of even this single passage here. Suffice it to say that Bartholomaeus's description became indelible from Bohemia's image for centuries to come, especially in England, where John Speed's and Stephen Batman's encyclopedias (Speed 1668; Bartholomaeus Anglicus and Batman 1584), repeating essentially the same information, were printed and reprinted after Trevisa's first translation of *De proprietatibus rerum* (Trevisa and Bartholomaeus Anglicus 1988; Seymour 1992). The work also enjoyed tremendous success in Jehan Corbichon's French translation, which was then readily used in massive volumes of sermon aids (Van den Abeele 1994; Binkley 1997;

Fig. 7.2 *Bonacus*. Illumination from *Der naturen bloeme* by Jacob van Maerlant between ca 1340–1350 (The Hague, Koninklijke Bibliotheek, KB KA 16, fol. 47v. Image in the public domain). http://manuscripts.kb.nl/zoom/BYVANCKB%3Amimi_ka16%3A047v_min_b1

Berlioz and Polo de Beaulieu 1994), thus, carrying the news of the Bohemian animals into a domain no longer accessible to us: oral culture. In addition, probably through the mediation of Corbichon's translation and/or the sermon exempla, the encyclopedic text also found its way to a small group of lavishly illustrated manuscripts, the *Livre des merveilles du monde* or *Secretz de la Nature* by an anonymous compiler. The four known manuscripts of this illustrated treatise were copied between 1427 and 1485 and treat countries and regions arranged alphabetically, mostly characterized by a marvel each. The unnamed beast appears in the description of the kingdom of Bohemia and is accompanied with the only known medieval illustration of the front-shooting variety (Beaugendre 1996, 18). The earliest known copy of the *Merveilles*, Pierpont Morgan Library, New York, 461 (France, possibly Angers, ca 1460), ends abruptly on fol. 27r with an unfinished entry on *Ethyopia*. Later copies include BnF, Paris, Département des Manuscrits, MS Français 1377–1379 (1427), vol. 1, fols 1r-27r; and MS 22971 (ca 1480). The incomplete manuscript marked as part of a private collection (Collection privée de la famille Charnacé, Paris), produced between 1450–1460, was formerly in the collection of Jean de Charnacé, Chateau d'Aulnoy, Coulommiers, Seine-et-Marne, but its present whereabouts are unknown (Friedman 2005).

Amid all this merging of information, the concept of the animal seems to have reached beyond its capacity and split into various different animals. At around the time of Bartholomaeus Anglicus's encyclopedia, two other influential *compendia* were in the making which essentially presented the Aristotelian/Solinian animal in various geographical places and with different wordings—in four somewhat chaotic entries. Albertus Magnus's *De animalibus libri XXVI* consists of three parts. Books 1–19 are based on Aristotle's works on natural science with some material coming from *De animalibus* by Ibn Sina (Avicenna, 980–1037), both translated from Arabic by Michael Scot. Books 20–21 seem to be Albert's own additions to Aristotle. The rest is a bestiary which, although unacknowledged, is based on *De natura rerum* by Thomas of Cantimpré (Albertus Magnus 1999a, 1:40; Egerton 2003, 89). Albertus mentions the *bonachus* and similar animals with or without names in various contexts across this hotchpotch of an encyclopedia. The Aristotelian part has a chapter dealing with the utility of horns in which Albertus describes an unnamed "wild ox," clearly based on Aristotle's *bonasos/monapos*, but reimagining the projectile and its effects in a unique manner, and not found anywhere else in the Latin texts.

> Book 12, chapter 7. [...] *Est autem quoddam genus agrestis vaccae sive bovis, quod cornua habet ad se invitem recurva non ad pugnandum apta: et illi dedit natura ut stercus longe et cum impetu posterius eiciat in oculos canum et venatorum se sequentium et eicit ipsum valde remote, et est viseosum, quod non de facili abstergitur: [et ideo] abstergendo ipsum detinentur canes et venatores insequentes ipsum.*
>
> There is a certain genus of wild cow (that is ox) which has horns that bend back in on themselves and are not suited for fighting. Nature gave this one the power of casting its dung to the rear, far and forcefully. Into the eyes of dogs and hunters following it. It casts it very far off and it is sticky so that it is not easily wiped off. The dogs and hunters following it are thus detained by wiping it off. (Albertus Magnus 1916, 889–890 and 1999b, 2: 981)

While the meaning is different—the excrement sticky rather than burning—the phrasing evokes the Aristotelian *bonasos* which "scorches the hounds' coats so that they rub off" (Aristotle 1991, 3:390). Solinus's translation of Aristotle in this instance was clearly superseded by the newly available translation arriving to Latin writers through the mediation of Arabic. The resulting discrepancy between the nature and effects of the excrement was not enough to pull a new animal out from the drawer, but it certainly added to the diversity of descriptions.

Albertus also includes similar animals elsewhere in *De animalibus*, namely in Tract 2, "On quadrupeds," in Book 22 (Albertus Magnus 1916, 1361, 1400, 1428). If Kitchell and Resnick is correct about his source and the direction of the transmission, then this is, in fact, Thomas of Cantimpré speaking. The two texts are nearly identical and—as opposed to the aforementioned sticky beast—they follow the bestiary tradition's Solinian phrasing and add a brief moralizing sentence:

> *Bonachum animal est, ut dicit Solinus, cui taurinum caput, corpus et iuba equina, cornua autem it multiplici flexu in se recurvata, ut si quis in ea offenderit, non vulneretur. Hoc animal cum in fugam vertitur, fluxu ventris fimum emittit post se spatium iugeris unius, cuius ardor quem attigerit adurit. Hiis armis submovet insequentes se. Signat hoc animal bonos prelatos, qui vite austeritatem quasi cornua in se habentes, dum eam in subditis exercent, subditos minime ledunt, quia in se quod hortendur ostendunt.*
>
> The *bonachum* is an animal, says Solinus, which has a bull's head and the body and mane of a horse, its horns are twisted manifold curving back onto

> themselves, so that if anyone attacks it is not harmed. When in flight, this animal discharges dung flowing from its stomach over an area of one *iugerum* behind itself, whose heat sets fire to anything it touches. These weapons ward off those who pursue it. This animal signifies good priests, who uphold an austerity of life for themselves, just as the [those] horns, while they practice that in submission, they hurt their subordinates minimally, because they reveal themselves through means of encouraging. (Thomas of Cantimpré 1973, 113)

In addition to the *bonachum*, the *henichires* (*henychiros*, *benitheres*) and the *zubrones* also appear as separate animals in the same book, rephrasing the projectile defense in different formulations. The *henichires*, according to Thomas, lives in the East and "discharges excrement at a distance of four paces when in despair" (*cum debilitatum fuerit, pugnat et in pugna stercus eicit per quatuor passus*). Note how the phrase "*cum debilitatum*" echoes the timid Aristotelian *bonasos* which defends itself by kicking and by voiding dung at the hunters "when it has lost strength" (Aristotle 1991, 3:391). The Bohemian *zubr* is described as an animal so fast that after discharging faeces from its rear end, it can turn in a flash and catch the dung on its horns before it reaches the ground (Thomas of Cantimpré 1973, 133). Considering the direction of the projectile, this version is halfway between the frontal effluent of Gervase's onager and Bartholomaeus's Bohemian ox, since it turns around to toss its own feces with its horns (Thomas of Cantimpré 1973, 172). Albertus Magnus's version omits Bohemia and places the animal in the forests of the north ("*in silvis aquilonaribus*" in Albertus Magnus 1916, 1428).

Amid all the splitting, new merging of animal concepts also takes place here: Thomas explains that the Indian bull shoots excrement at dogs and hunters "*quoniam eiciunt stercus post se remotissime, cuius calore et fetore retardant canes et venatores*" (Thomas of Cantimpré 1973, 166). The Indian bull is found in most Second Family bestiaries, where, instead of protecting itself with his excrement, it breathes fiercely to avoid being tamed if captured. This, in turn, is based on 52.36 in the *Collectanea*: "Indian bulls [...] twist their horns around with all the suppleness they want. By the hardness of their hide they turn aside every spear and they are so filled with indomitable ferocity that when they are captured they abandon breathing. (*Indicis tauris. [...] Hi quoque circumferunt cornua flexibilitate qua volunt. Tergi duritie omne telum respuunt et tam inmiti feritate, ut capti animas proiciant furore*; Solinus 1895, 189). In fact, the rare phrase "*anima proiciant*" goes back to Virgil's *Aeneid* 6.436, where it translates as "they flung away their lives" (Virgil 1916, 563). Thomas, however, clearly ignores this and borrows the *bonnacon*'s weaponry for his description.

It seems that after their proliferation in thirteenth-century western encyclopedias, the beast(s) appear(s) in texts produced closer and closer to its purported habitat in Eastern Europe, written in or about Bohemia itself. Giovanni Marignoli, Emperor Charles IV's chaplain who appears to have accompanied the emperor to Prague in 1354–1355, for example, incorporates Bartholomaeus's passage into his rambling *Chronicle of Bohemia*, changing only a tiny detail: while Bartholomaeus's beast "*habet magna cornua & ampla, cum quibus se defendit*," Marignoli's has "*magnis cornibus, cum quibus tamen se* non *defendit*" going back to the (rather ambiguous) Solinian description: "curling back on themselves in such a way that if anyone attacks it, it is not harmed." Also note the use of the verb '*proicit*' which certainly moved far away from the original Virgilian phrase used by Solinus and is used in the literal sense as "shoot" or "fling forward."

> *Bestie eciam domestic & silvestres innumeriablis, sicut ursi, apri, cervi, caprelle, tragelaphi, bubali, bisontes. Est & ibi bestia habens magnitudinem bovis, ferox & seua cum magnis cornibus, cum quibus tamen se non defendit. Hec bestia habet sub mento folliculum magnum, in quo aquas colligit multas & currendo aquas calefacit miro modo in illo folliculo, quas super canes & venatores proicit, & quicquid tetigerit, depilat, & urit horribiliter.*

> There are also countless domestic and woodland animals such as bears, boar, deer, wild goats, gazelles and bison. And among these beasts is a [beast] which is the size of an ox, fierce, wild, and has big horns, with which it does not defend himself. This animal has under its chin a large bladder in which it collects much water and when it runs, it warms the water in its bladder in a marvellous way, which it then shoots at hunters or dogs and horribly removes hair from and scalds whatever it touches. (Marignoli 1768, 138–39)

Thomas's dung-flinging *zubr* is also of interest here because it is said to "live in Bohemia and is called *thurones* in Poland." At about the same time as Marignoli's chronicle, the *Glossarius*, written under the direction of Claretus de Solentia (Bartholomaeus of Chlumec), officially added the animal to the natural history of medieval Bohemia. The *Glossarius* was conceived to translate nearly 7000 foreign words into Czech. For forest dwelling cattle, Klaret's source was almost certainly Thomas of Cantimpré's *Liber de natura rerum*, or its German translation (Meduna and Kyselý 2009, 242). The nomenclature is so superbly mixed up (and partially made up) that it cannot be used as evidence for anything but Claretus's creativity and breadth of classical erudition. For Claretus, the *bubalus* is identical with the *thur*, even though they are two different animals in the *Liber*. Thomas's other name for the *bubalus* (*bissontes*) is translated separately as *walda*. He goes on to translate *tigris* as *zubr*, turning around Thomas's statement that the *zubr* is erroneously called *tygris* in Latin: "*falsoque obtinuerunt in illis regionibus a Latinis nomen tygris, et hoc propter velocitatem*" (Thomas of Cantimpré 1973, 172). Finally, the Czech equivalent of the *banochus* is *dobranyk*. As Šedinová explains, this translation, again, is proof of Claretus's creative *modus operandi*: in a truly Isidorean spirit, there are "names borrowed from Thomas' encyclopaedia to which Claretus appended a Czech equivalent lacking any connection to Thomas' descriptions, but rather influenced by the form of the Latin term," such as the pair *dobranyk—banochus* (cf. the Czech adjective *dobrý*, "*bonus*" (Šedinová 2016, 117).

Returning from this short expedition in Eastern Europe, the first question is naturally whether the *bonasus* and the Eastern European beast is the same animal. Or rather: was it perceived as the same animal when more than one version was available for the reader? While, given their extremely different habitats, the real animals would certainly not be identical even if cognate, the textual animal can indeed be seen as an evolving amalgamation of various concepts revolving around the figure of large, fierce bovines, whether in Asia or Eastern Europe. The unquestionable textual connections, discernible through the analysis of this single passage by the four contemporaneous authors, demonstrate that in this instance much of the knowledge transfer took place outside the texts: besides Albertus's "borrowing" from Thomas of Cantimpré, it is hard to detect verbatim borrowing, no matter how striking the textual similarities (and discrepancies) are. It is more justifiable to presume that the circulation of information took place in mendicant/missionary literacy and university/*studium* circles which found their way into works of science.

Since the latter authors were all active in or connected to Europe's north-eastern provinces, it is tempting to suggest that they actually saw European bison or aurochs with their own eyes and adapted received encyclopedic knowledge to fit their own personal experience. Michael C. Seymour (1992, 161) notes to *De proprietatibus rerum*, for example, suggest that his "account of Bohemia [including the fierce beast] possibly derives from Bartholomaeus Anglicus's personal knowledge and 'Erodatus' [unidentified geographical author in the thirteenth century]." Banks and Binns say about the *Otia imperialia* that "it is clear from Gervase's description of Poland in Book II that he had had a first-hand description of that country from a native of Poland, this information on the Polish onagers may have come from the same source" (Gervase of Tilbury 2002, 684). However, these suggestions remain tenuous based on evidence that can only be circumstantial given our present knowledge of their exact sources, their biography, and the place and time of writing.

The Tweakings

There is still not a single real animal after all this reading, only faint reflections of Aristotle's *bonasos* in the pieces of a broken mirror. What can be seen from the overview above is the glacial transformation of the concept of an animal in an extremely varied corpus across time and space. Updating or changing the passage was clearly not on the agenda for bestiary scribes. Later, it was rather conservatively performed by encyclopedists whose work retained patches from the *bonasus* story, applying its motifs to "new" animals, that in their turn came to be frozen in time once again—Bartholomaeus's and Thomas's descriptions were handed down unchanged for centuries. The medieval authors closest to the *bonasus*'s alleged fire, and even early modern naturalists who had etchings, bones, and eyewitness accounts at their disposal, regurgitated information that essentially came from the Aristotle-Pliny-Solinus trifecta. At the same time, the texts are not stagnant in every respect. While the concept of the animal hardly changes except for the increasingly chaotic nomenclature, the variation of the wording is surprising and remains creative across the textual transmission. Burrowing deeper into the text, there are instances of philological tweaking that deserve a brief mention and more detailed consideration in the future. While the main characteristics of the *bonnacon* (and friends) seems relatively constant across a wide variety of texts, which in itself is tremendously interesting, a little more spadework will reveal more about this odd beast in the making.

Firstly, the question of the projectile. As quoted previously, Aristotle, Pliny, and Solinus all squarely said that the *bonasus* shot dung at its pursuers. I have already dealt with the difference between the coat of the dogs being rubbed off by hot dung (Aristotle) and having to rub off sticky dung (Albertus/Thomas of Cantimpré) as a potential translation issue, but it was still dung. The *bonnacon* in the Second Family bestiary tradition, clearly the same animal, however, ejects noxious fumes. Anyone who has experienced how treacherous minims can be in medieval Latin manuscripts (or simply has dyslexia), will immediately realize that this is probably the result of misreading. A minim is a single written stroke: a single minim is the letter "i," two minims could stand for "u" or "n," while three minims could be anything from "m," "ui," "iu," "ni," to "in," so it is probable that the nine minims of *fimum* were at one point copied as ten minims of *fumum*. *Fimum*, the accusative of masculine noun *fimus*, *fimi* (or neutral *fimum*, *fimi;* dung), became *fumum*, the accusative of masculine noun *fumus*, *fumi* (fume). The dung-projectile, "*ventris fimum egerit*" in Solinus became noxious fumes, "*ventris fumum egerit*," in medieval bestiary manuscripts (for example, London, British Library, MS Add. 11283, fol. 5v, England, ca 1170). Notably, this is not the case in Third Family bestiaries, which retain the original "*fimum*" wording (Dines 2008, 177). This small detail not only changes the idea of the animal and the accompanying images but can be used to learn more about the direction and sequence of transmission when it comes to Second Family Bestiaries. The difference is not always apparent in the visual representations. Faithful to the text, in the Second Family bestiaries, some images suggest fumes, either by depicting gaseous-looking emissions, no emissions, or only hunters covering their noses: Kongelige Bibliotek, Gl. kgl. S. 1633 4°, fol. 10r, Bibliothèque Nationale de France, lat. 3630, fol. 78r; Bodleian Library, Bodleian Library, MS. Bodley 533, fol. 5r; Bodleian Library, MS. Bodley 764, fol. 16r; British Library, Harley MS 4751, fol. 11r; British Library, MS Add. 11283 , fol. 5v. Some emit an ambiguous substance: Museum Meermanno, MMW, 10 B 25, fol. 8r; Morgan Library, MS M.81, fol. 37r; MS. Ashmole 1511, Fol. 18r; Bodleian Library, MS. Douce 88, fol. 10r; British Library, Royal MS 12 F. xiii, fol. 16r—noting that the reddish brown color shown in this picture is not necessarily a marker of excrement but may just stand for the heat of the fumes (Fig. 7.3). At the same time, despite the obvious discrepancy between text and image, the antler-bearing animal in Second Family Koninklijke Bibliotheek, KB, KA 16, fol. 47v is unmistakably releasing solid droppings (cf Fig. 7.1). In the Third Family bestiary, London, Westminster Abbey MS 22, fols 25rv, where the text clearly indicates "*fimum*," the animal releases what looks like a smallish deluge of downward pouring liquid hitting the shield of a solitary hunter. Bodleian Library, MS. e Mus. 136, fol.

Fig. 7.3 Bonacon in the *Rochester Bestiary*, British Library, Royal MS 12 F. xiii, fol. 16r. (Image in the public domain). https://commons.wikimedia.org/wiki/File:RochesterBestiary_detail_Bonnacon.jpg

24v looks like a poorer version of the Westminster image, where the projectile ends up like a second, shorter tail. Finally, truest of all to the "*fimum*" version, in the Third Family MS, Oxford, Bodleian Library MS Douce 88, fol. 92v, the image definitely shows release of dark brown solids.

Besides variants resulting from misinterpreting or miscopying the words and phrases used for the same phenomenon, for example, the dung (where applicable), the effects of the various projectiles on dogs and hunters, and the description of the horns, all show the unique personal mark of the medieval author reusing the text. As shown previously, the differences between the descriptions of Pliny and Solinus are so readily discernible that they allow identifying the latter as the sources of the *bonnacon* descriptions on maps and bestiaries. With the arrival of Aristotelian texts and other sources, the wording became more varied. For example, although *fimus* became *fumus* in bestiaries, Albertus Magnus and Thomas of Cantimpré clearly preferred Aristotle's version in Michael Scot's translation, and reformulated the idea using *fimus* and its synonym, *stercus*, in multiple variations.

Intertextual variations become fuzzy at this point. While about the *bonachum* Thomas says "*cum in fugam vertitur, fluxu ventris fimum emittit post se spatium iugeris unius, cuius ardor quem attigerit adurit*" (Thomas of Cantimpré 1973, 113), the oriental *henichires*'s identical mode of defense is described about a dozen pages later as "and in fight it throws manure at a distance of four paces and it is due to fear" ("*et in pugna stercus eicit per quatuor passus, et hoc causa timoris*" in Thomas of Cantimpré 1973, 133). This is close to Albertus Magnus's *bonacus*, about which he says "when it weakens, it fights and throws manure at the hunters at a distance of four paces: it is released out of fear" ("*Cum vero debilitatur, pugnat et contra venatorem eicit stercus ad quatuor passus: solvitur enim ex timore*" in Albertus Magnus 1916, 658). More yet, in the description of the Indian bull, Thomas (or his source) ignores Solinus's *taurus indicus* (Solinus 1895, 189), and echoes Aristotle's entry on the *bonasos* instead, saying "they throw manure far behind themselves, by whose heat and stench they hold up dogs and hunters lest they come near to them" ("*quoniam eiciunt stercus post se remotissime, cuius calore et fetore retardant canes et venatores, ne eis approximant*" in Thomas of Cantimpré 1973, 166). The use of *stercus*, instead of *fumum* (or *fimum*) clearly indicates that the author had access to Aristotle's *bonasos* outside of the Pliny-Solinus-Physiologus transmission: as noted, it is based on a post-Scotus Aristotelian concept of the animal and its discharge.

The wide range of words describing the scorching effect of the projectile across the sources also suggests a dynamic web of concepts and seemingly conscious creative wording. Besides the usual variant readings (e.g. *exurit* -> *et urit*), all the authors seem to have been keen to put their personal mark on the received material in more subtle ways than appending additional sentences of interpretation or moralization. Solinus's "the heat of which burns anything it touches. In this way, it keeps away pursuers with its harmful excrement" is phrased very similarly in Thomas of Cantimpré, and rephrased and expanded to a greater extent as "horribly removes hair from and scalds whatever it touches" by Bartholomaeus Anglicus and by Marignoli: "*cuius ardor quicquid attigerit amburit: ita egerie noxia submovet insequentes*" (Solinus 1895, 168); "*cuius ardor quem attigerit adurit. Hiis armis submovet insequentes se*" (Thomas of Cantimpré 1973, 113); "*quicquid tetigerit depilat horribiliter & exurit*" (Bartholomaeus Anglicus 1964, 639); "*quicquid tetigerit depilat et urit horribiliter*" (Marignoli 1768, 138). In another context, in Bersuire's Dominican sermon aid, the same text is inflated into a judiciary exemplum "and opening its mouth on account of impatience and impropriety, it throws the water of its wickedness over them [...] and so the dogs and the hunter, that is the judge and the prosecutor, fearing them withdraw their charge and pardon the unimproved man" ("*et aquam malitiarum suarum aperto ore per impatientiam et improperiam proiicit super illos [...] et sic canes et venator id est iudex et accusator timentes eius impetum retrocedunt et eum incorrectum dimittunt*" (Bersuire 1583, 614).

The lines describing the horns also vary from one text to another, sometimes completely changing the meaning. Pliny's "horns bent inwards upon each other" (*cornibus ita in se flexis*) is reasonably similar to Solinus's "horns twisted manifold, curling back on themselves" (*cornua autem ita multiplici flexu in se recurrentia*), and Cantimpré's "twisted manifold, curving back onto themselves" and "its horns are very much twisted and useful for fighting" ("*multiplici flexu in se recurvata*," and "*et cornua eius multum recurvata et ad pugnandum habilia*" s.v. *bonachum* and *henichires* in Thomas of Cantimpré 1973, 113 and 133, respectively). However, as noted, the description of the horns of what may or may not be understood as the same animal in Bartholomaeus Anglicus ("*magna cornua & ampla*": long and wide) and Gervase of Tilbury ("*ad modum damarum*": like an antler) truly sets their versions apart.

These few examples suffice to demonstrate that even a few variables can (and did) produce a surprisingly large pool of potential combinations of sources and rewriting. These are, of course, tremendously valuable in manuscript studies, since they can be used as evidence for the direction of intertextual relationships, which inform conclusions about dating and even author biographies. As far as the textual animal is concerned, however, they also provide an opportunity to further stress the contradiction between the conservative treatment of the content of the animal's description, and the creative mind of the medieval scribes, authors, and illustrators confidently tweaking the phrases and words used and adding explanatory phrases where they felt it was necessary. About the former—painstaking manuscript studies—Debra Higgs Strickland points out that "the repeated tweakings of categories privilege manuscript classification over the richer cultural meanings generated by the combined impact of texts and pictorial images" (Higgs Strickland 2011, 816). Although, as briefly shown above, understanding the relationship between manuscripts adds to our understanding of the *evolution* of said texts and images, something not to be underestimated, the combined impact of texts and pictorial images is indeed something to keep an eye on, especially if some readers (and authors) had more than one version of the same description at hand, and did not hesitate to forge their own ideas about the animals in question.

The Meanings

But why did they bother at all? The key, as it often is in the study of medieval texts, lies in the idea of the four levels of exegesis—the literal, allegorical, moral and anagogical—expounded in Hugh of St Victor's *Didascalicon*, composed in the late 1130s. As Umberto Eco explains:

> this is the project taken up by bestiaries, the *lapidaria*, the *imagines mundi*, all formed on the Hellenistic model of the *Physiologus*. Certain properties are predicated of every animal, plant, part of the world, or even in nature, and, on the basis of an identity between one of these properties and one of the properties of the supernatural being that is to be metaphorized, a correlation is established. A network of cultural information exists, which functions as a cosmological code. (Eco 1983, 235)

This takes us to an important juncture in the discussion which returns to what Debra Higgs Strickland (2011, 816) called "rich cultural meanings" and Eco (1983, 236) places in "a maze-like network of cultural properties." While there is no room here to comprehensively deconstruct this metaphorical additional layer across the *bonasus/bonnacon* passages, the moral meaning of the existence of various animals confirmed by every single animal description—whether using explicit analogies or allegories or not—cannot be ignored in this overview. For example, the reason why Bersuire's entry is more detailed about the pursuit element than his sources, Gervase and Bartholomaeus Anglicus, is because he builds his exemplum around it. The text easily lent itself to moral additions and biblical parallels, because it was always more than just a description of an animal. This was clear for medieval readers even without added moralization.

On this level, the present summary of tiny lexical details makes it clear that what connects the web of seemingly different—but textually/conceptually similar—beasts is that flight is their primary mode of defense and that they only appear as prey in a human context, specifically in the context of hunting, never fleeing from non-human animals (except for the humans' dogs). They are hunted but they win. While images of hunting place human agency into animal descriptions, it is important to remember that the real humans in these stories are the *bonnacon* itself. The animals' resourcefulness and natural characteristics are human, the pursuers are not. Various examples have been noted above, for instance, the four Third Family bestiaries, which added a moralizing sentence from the Scripture likening impudent people to the animal. Before Bersuire appropriated the Bohemian beast for his moralizing *Reductorium*, Gervase of Tilbury already placed the "Polish onager" in a context that reached beyond the alleged Polish reality of the animal, saying (as previously mentioned) "the psalm in its literal meaning says of these: 'The wild asses shall expect in their thirst'" (Gervase of Tilbury 2002, 684–85). Even though the "*expectabunt*" probably harkens back to the odd behavior of Aristotle's *bonasos* which "stands to meet" the hunters, Gervase reaches for the Bible to fill this detail with meaning, and connect it with the water element in his description. For the same reason, but conveying a different message, Thomas of Cantimpré added that the *bonachum*, because of its odd horns, signifies a good priest who is stern with himself in his own life, but does not hurt his followers (Thomas of Cantimpré 1973, 113).

In conclusion, the Paeonian/Phrygian/Polish/Bohemian animal crops up in an increasingly chaotic body of different medieval texts, updated, especially after the rediscovery of Aristotle, by nearly undetectable rephrasing, splitting and merging, and cross-pollinating between texts and animal concepts. A number of instances have been shown to illustrate the dynamic but rather oblique intertextual bonds that created the concept of this unique animal across time. The mobility of the passages in different genres (such as Gervase of Tilbury's and Bartholomaeus Anglicus's beast used in Bersuire's exempla, or Corbichon's translation of Bartholomaeus's *De proprietatibus rerum* used in the *Merveilles du monde* and even Jean Mansel's fifteenth-century chronicle, *La fleur des histoires*), also attesting to the metaphorical importance of the beast, deserves further research.

As Kellie Robertson points out, "[w]hereas [C. S.] Lewis had assumed a "trickle-down" model of scholastic knowledge into extra-academic writing, a more circulatory model now predominates in medieval literary studies [...]. A more dynamic knowledge production that is discernible in the natural problems [...] as they move from academic registers to more popular ones and back again, often showing signs of having been modified by their contact with new ideas in these extra-academic settings" (Robertson 2017, 10). Importantly, the story of the *bonasus/bonnacon*, tossed back and forth in the process of knowledge production, points to a significant methodological observation, which has wider implications in interpreting medieval texts about "odd beasts" or "mythical hybrids." Given the variation in wording, the additions, and the multiplication of curiously similar animals, the ultimate question here remains whether the animals were genuinely connected, even identified, in the minds of medieval and early modern readers? Our evidence for the readers' interpretation of the texts comes from writers and scribes who reused received information. Even this brief analysis of the reuse, afterlife, and variants of the descriptions shows that they were indeed connected in unexpected ways. Together they form a maze-like network of cultural properties: the nebulous concept of the large bull-like animal with a projectile defense system that can be used in myriad ways to contemplate human nature, especially because it features something so universally relatable—commonly known as bullshit.

References

Albertus Magnus 1916. *Albertus Magnus de Animalibus Libri XXVI. Nach Der Kölner Urschrift*, edited by Hermann Stabler. 15. Münster: Aschendorff.

Albertus Magnus 1999a. *On Animals: A Medieval Summa Zoologica* Vol. 1, edited by Kenneth F. Kitchell, and Irven Michael Resnick. Foundations of Natural History. Baltimore MD: Johns Hopkins University Press.

Albertus Magnus 1999b. *On Animals: A Medieval Summa Zoologica* Vol. 2, edited by Kenneth F. Kitchell, and Irven Michael Resnick. Foundations of Natural History. Baltimore MD: Johns Hopkins University Press.

Aristotle. 1991. *History of Animals* Vol. 3, translated by David M. Balme. Loeb Classical Library 439. Cambridge MA: Harvard University Press.

Bartholomaeus Anglicus. 1964. *De Rerum Proprietatibus*. Frankfurt: Wolfgang Richter, 1601. Frankfurt am Main: Minerva.

Bartholomaeus Anglicus, and Stephen Batman. 1584. *Batman Vppon Bartholome His Booke De Proprietatibus Rerum, Newly Corrected, Enlarged and Amended: With Such Additions as Are Requisite, Vnto Euery Seuerall Booke: Taken Foorth of the Most Approued Authors, the like Heretofore Not Translated in English.*

Profitable for All Estates, as Well for the Benefite of the Mind as the Bodie. London: Thomas East.

Baxter, Ron. 1998. *Bestiaries and Their Users in the Middle Ages*. Stroud–London: Sutton Publishing–Courtauld Institute.

Beaugendre, Anne-Caroline, ed. 1996. *Les merveilles du monde ou Les secrets de l'histoire naturelle*. Arcueil: Anthèse.

Berlioz, Jacques, and Marie-Anne Polo de Beaulieu. "Les Recueils d'exempla et La Diffusion de l'encyclopédisme Médiéval." In *L'enciclopedismo Medievale: Atti Del Convegno San Gimignano, 8–10 Ottobre 1992*, edited by Michelangelo Picone, 179–212. Ravenna: Longo, 1994.

Bersuire, Pierre. 1583. *Reductorium morale super totam bibliam.*. Venice: apud haered. Hieronymi Scoti.

Binkley, Peter. "Preachers' Responses to Thirteenth-Century Encyclopaedism." In *Pre-Modern Encyclopaedic Texts: Proceedings of the Second COMERS Congress, Groningen, 1–4 July 1996*, edited by Peter Binkley, 75–88. Brill's Studies in Intellectual History 79. Leiden–New York: Brill, 1997.

Buffon, Georges Louis Leclerc de, and James Smith Barr. 1807. *Natural History: Containing a Theory of the Earth, a General History of Man, of the Brute Creation, and of Vegetables, Minerals, &c.* Vol. 8. London: H. D. Symonds.

Clark, Willene B. 2006. *A Medieval Book of Beasts: The Second-Family Bestiary: Commentary, Art, Text and Translation*. Woodbridge: Boydell.

Crane, Susan. 2013. *Animal Encounters: Contacts and Concepts in Medieval Britain*. 1st ed. Middle Ages Series. Philadelphia PA: University of Pennsylvania Press.

Dines, Ilya. (2008). Ilya Dines, "*A Critical Edition of the Bestiaries of the Third Family*." PhD diss., The Hebrew University of Jerusalem, 2008.

Dines, Ilya. "The Problem of the Transitional Family of Bestiaries." *Reinardus* 24 (2012). 29–52.

Eco, Umberto. "The Scandal of Metaphor: Metaphorology and Semiotics." Translated by Christopher Paci. *Poetics Today* 4, no. 2 (1983). 217.

Egerton, Frank N. "A History of the Ecological Sciences, Part 9. Albertus Magnus: A Scholastic Naturalist." *Bulletin of the Ecological Society of America* 84, no. 2. (2003). 87–91.

Friedman, John. "Robinet Testard, Court Illuminator: His manuscripts and his debt to the graphic arts." *Journal of the Early Book Society* 8 (2005). 152–96.

Gervase of Tilbury. 2002. *Otia Imperialia : Recreation for an Emperor*, edited by Shelagh E. Banks, and James W. Binns. Oxford Medieval Texts. Oxford: Clarendon Press.

Gesner, Conrad, Jacobus Carronus, Pierre Belon, and Gulielmus Rondeletius. 1551. *C. Gesneri Historiæ Animalium: Liber I Qui Est de Quadrupedibus Viviparis; Liber II. de Quadrupedibus Oviparis; Liber III. Qui Est de Avium Natura; Liber IIII. de Piscium & Aquatilium Animantium Natura; Liber*

V. Qui Est de Serpentium Natura Ex Variis Schedis. 5 vols. Tiguri–Zürich: Christoffel Froschauer.

Higgs Strickland, Debra. "Review of 'From the Ark to the Pulpit: An Edition and Translation of the 'Transitional' Northumberland Bestiary (13th Century)' by Cynthia White." *Speculum* 86, no. 3 (2011). 816–817.

Jones, Richard. 2013. *The Medieval Natural World*. London: Routledge.

Leuphana Universität Lüneburg. n.d. "HyperImage: EbsKart." Accessed August 3, 2018. http://www2.leuphana.de/ebskart/.

Marignoli, Giovanni. "Chronicon Bohemorum." In *Monumenta Historica Boemiae: Nusquam Antehac Edita, Quibus Non Modo Patriae, Aliarumque Vicinarum Regionum, Sed et Remotissimarum Gentium Historia Mirum Quantum Illustratur : Collegit, et Partim Ex Autographis, Partim Ex Legitimis Apographis Codicibus Recensuit, ... Denique Locupletissimo Indice Instruxit*, edited by Gelasius Dobner. 279–288. Prague: Clauser, 1768.

McCulloch, Florence. 1962. *Mediaeval Latin and French Bestiaries*. Chapel Hill: University of North Carolina Press.

Meduna, Petr, and René Kyselý. "O Zvířeti Velkém Jako Slon, Mezi Jehož Rohy Si Mohou Sednout Tři Muži: Pratur ve Středověku Čech a Moravy-Historická a Archeozoologická Analýza [About an Elephant-Sized Animal That Could Seat Three Men between Its Horns: The Aurochs in Medieval Bohemia and Moravia; A Historical and Archaeozoological Analysis]." *Památky Archeologické* 100 (2009). 241–260.

Oppenraay, Aafke M. I. van. "Michael Scot's Translation of Aristotle's *Books on Animals* and the Pleasures of Knowledge." *Quaestio* 15, January (2015). 413–22.

Pliny the Elder. 1855. *The Natural History of Pliny*. Translated by John Bostock, and Henry T. Riley. London: Bohn.

Robertson, Kellie. 2017. *Nature Speaks: Medieval Literature and Aristotelian Philosophy*. 1st edition. The Middle Ages Series. Philadelphia PA: University of Pennsylvania Press.

Šedinová, Hana. "Aristotle's Carp as Claretus' Bird 'Comor'? Tracing the Origin of One Medieval Term." *Auc Philologica Graecolatina Pragensia* 2016, no 2 (2016). 111–123.

Seymour, Michael C. 1992. *Bartholomaeus Anglicus and His Encyclopedia*. Aldershot: Variorum; Brookfield VT: Ashgate Pub. Co.

Solinus, C. Julius. 1895. *Collectanea rerum memorabilium*, edited by Theodor Mommsen. Berlin: Weidmann.

Speed, John. 1668. *A Prospect of the Most Famous Parts of the World*. London: Roger Rea.

Stewart, Patricia. 2012. *The Medieval Bestiary and Its Textual Tradition*. Fife: University of St Andrews.

Thomas of Cantimpré. 1973. *Liber de Natura Rerum: Editio Princeps Secundum Codices Manuscriptum 1: Text*. Berlin: Walter de Gruyter.

Trevisa, John, and Bartholomaeus Anglicus. 1988. *On the Properties of Things : John Trevisa's Translation of Bartholomaeus Anglicus De Proprietatibus Rerum, Vol. 3, A Critical Text*, edited by Michael C. Seymour, and Gabriel M. Liegey. Oxford [u.a.]: Clarendon Press.

Van den Abeele, Baudouin. "Bestiaires Encyclopédiques Moralisés: Quelques Succédanés de Thomas de Cantimpré et de Barthélemy l'Anglais." *Reinardus: Yearbook of the International Reynard Society* 7 (1994). 209–228.

Virgil, 1916. *Aeneid.* Translated by Henry Rushton Fairclough and George Patrick Goold. Cambridge, MA: Harvard University Press. https://doi.org/10.4159/DLCL.virgil-aeneid.1916.

Werness, Hope B. 2007. *The Continuum Encyclopedia of Animal Symbolism in Art.* London: Continuum.

Westrem, Scott D. 2001. *The Hereford Map: A Transcription and Translation of the Legends; with Commentary.* Terrarum Orbis, History of the Representation of Space in Text and Image 1. Turnhout: Brepols.

Wotton, Edward. 1552. *Edoardi VVottoni Oxoniensis De Differentiis Animalium Libri Decem .. / Deinde Omnium Animalium Nomenclaturae, Itémque Si.* Paris: Michel de Vascosan.

CHAPTER 8

What's in a Noun? A Short *Caveat* Regarding the Difficulties of Identifying Medieval Animals in Texts

Richard Trachsler

Introduction

One of the most aggravating problems when dealing with medieval texts mentioning animals is that we oftentimes just do not know what animal they actually refer to in terms of modern taxonomy. This uncertainty applies not only to imaginary creatures and fantastic beasts such as the siren or the centaur, the dragon or the unicorn, but to many "real" animals that are perfectly well identifiable today. Medieval authors can use a name that seems familiar to refer to a different animal, or, more commonly, describe an animal in a way that is at odds with modern concepts, so that we sometimes wonder which beast they actually had in mind. Rather than offering solutions, the present article will present a survey of the problems medievalists face when trying to figure out what kind of animal is actually being mentioned in a text. Mainly, this will result in the

R. Trachsler (✉)
Romanisches Seminar, Universität Zürich, Zürich, Switzerland
e-mail: richard.trachsler@uzh.ch

L. Bartosiewicz, A. M. Choyke (eds.), *Medieval Animals on the Move*, https://doi.org/10.1007/978-3-030-63888-7_8

following series of *caveats* based on concrete examples: Never take for granted that a medieval animal, despite its identical name, is identical to its modern equivalent; never take for granted that a medieval animal, despite a different name, does not have a modern equivalent, never take for granted that a translation or description given by a medieval writer is accurate, and, most importantly, always check why a medieval author might be offering that specific translation or description. The overall result might look disappointing, since it contains a strong *pars destruens*, but it also shows that almost every time, at least some things can be explained. In order to create a good argument, however, philologists must look at a wide range of texts and be prepared to leave their comfort zone.

Identifying Animals in the Written Record

The panther is a good example of an animal that occurs in virtually every bestiary and encyclopedia, but has little in common with the animal modern zoologists refer to by the name of *Panthera pardus*. *Panthera pardus* is the scientific name for the leopard, which, indeed, some modern languages also call "panther". Medieval writers, however, used these terms for two beasts that they considered to be unrelated: the medieval panther is a fantastic creature covered with spots of many colors, which has the ability to draw all other animals toward itself thanks to its mellifluous and charming voice. The panther is friends with all animals—hence its name *pan-thera*, "all animals"—the only exception being the dragon, who hides in its hole or attempts to shut its ears in order to avoid hearing the panther's beautiful voice. On this basis, the panther is often said to be a figure of Christ, who attracted men and women with his sweet speech. The iconographical tradition surrounding the animal is completely eclectic: the body of the panther can resemble anything from a dog to a horse. The only stable element is its multi-colored markings, consisting generally of blue and red circles, most frequently on a white background (Van den Abeele 2015). The medieval leopard is a different animal altogether (Trachsler 1992): most often, it looks like a lion and, most often, its fur does not have the characteristic spots that are systematically mentioned when the animal is described in texts. Its perception is almost entirely negative: the leopard is the adulterous offspring of the lioness and the *pard*, an unidentified spotted feline. This also explains its name: *leo-pard*. It is said to be fast, aggressive, cruel, and treacherous in didactic or religious literature, and is thus used as a figure of the devil (Trachsler 1992).

Clearly, neither the medieval panther nor the medieval leopard is identical to *Panthera pardus*, their modern zoological counterpart. And clearly they are two entirely different animals. On the other hand, since every single description of these two animals in medieval texts mentions their spotted fur, it is just as evident that there is some kind of connection not only between the *Panthera pardus* and its medieval counterparts, but also between the medieval panther and the medieval leopard. It is hardly a coincidence that one of the two creatures is entirely positive, whereas the other is predominantly negative. Medieval animals may not always conform to modern zoology, but they often form a system of their own. The animal, in the Middle Ages, always carries a semantic load, which influences its perception, that is, the way people see and, hence, name it.

Why begin this philological inquiry on the difficulty of identifying medieval animals with ideas that are little more than common knowledge? Because it is helpful to keep in mind that sometimes terminological "errors" can be induced by a variety of elements, ranging from etymology to real or imaginary qualities which a given animal might share with another. These considerations are particularly important whenever a writer or a copyist is confronted with an animal they have difficulty identifying.

Far-Away Animals Moving into the Record

Problems of identification will occur more frequently the greater the geographical and cultural distance that separates the writer from a given animal. Typically, terminological difficulties arise when exotic animals need to be described and named. This is nicely demonstrated by a striking example regarding the crocodile, a creature regularly depicted and described in the bestiary tradition, but whose physical appearance varies greatly in its illustrations. When the pilgrims in Egypt first saw crocodiles, they had no suitable animal from home they could compare it to. The best they could do was to say that, in the Nile, there are "*pessima animalia* [...] *ad similitudinem draconis*": "very evil animals, resembling dragons".

Speaking of the Nile, the Irish friar Symon Simeonis states (ca. 1323):

> *His autem fluvius posset in multis aliis a predictis commendari, nisi esset quod quedam animalia pessima resident in eo, que sunt ad similitudinem draconis. Devorant quidem equos et homines in flumine si inveniant, [et] super ripam devorare non postponunt. Et ista animalia cocatrix vulgariter appelantur.*

> Many other excellent things might be said about the river [Nile], were it not the retreat of a highly noxious animal, resembling the dragon, which devours both horses and men if it catches them in the water or on the banks. This animal is called by the people *cocatrix*. (Symon Simeonis 1960, 66–67)

The interesting element here is that an imaginary creature such as the dragon provides a familiar reference to picture an exotic but real beast. When trying to identify the unknown, we generally tend to compare it to what we know. Usually, though, writers would not resort to such extreme measures and, rather, preferred comparisons with entities pertaining to the familiar surroundings of their readers. An often mentioned example is Marco Polo (1998), who describes Eastern fauna using the entire spectrum of Western animals he has at his disposal (for the fauna he describes, see Levy 1997 and Faucon 1997).

To him, basically, crocodiles—the name occurs nowhere in his text—are big snakes with legs: *grans culueuvres et* [...] *serpens* "huge snakes and serpents", ten feet long, some more, some less, and thick as a barrel. And they have *.II. jambes devant, prez a la teste*, "two legs in front, near the head" and a mouth so enormous *que moult bien engloutiroit un homme entier* "it could easily swallow a whole man" (Marco Polo 2005, 78). A tiger —again, the name is not found in his work— is a lion with stripes: *lion tuit vergié du lonc de noir et de vermeil et de blanc*, "a lion striped in black and red and white, over the entire length of its body" (Marco Polo 2004, § 90, 6). Here, the modern scholar should avoid jumping to the conclusion that the text is dealing with snakes or lions. Rather, snakes and lions were simply the closest animals Marco Polo could think of when seeing crocodiles and tigers. We see and perceive the world around us according to what we are familiar with. On some occasions, our expectations are challenged by what we see: Famously, Marco Polo is the first to explain to his fellow Westerners after having seen a rhinoceros that a "unicorn"—this time, the word is used in his text—is not the pony- or goat-like animal they know from their books, but a terrifying creature, large as an elephant, with the head of a wild boar, feet of an elephant and hair of a buffalo. When faced with something unknown, in order to describe it, people often disassemble the strange entity into familiar fragments of knowledge: elephant, buffalo, or wild boar. But something more important can be gathered from Marco Polo's description of the unicorn when he concludes: *elle est mout laide beste a veoir. Il ne sunt pas einsi come nos de ça dion e deviçon*, "It is a very ugly beast to behold. They are not the way people at home say they are and describe them" (Marco Polo 1982, 544. See also the excellent note in Marco Polo 2009, 111). Books can be inaccurate when confronted with reality. In this specific instance, Marco Polo had the word to name the animal, but it simply was not adequate.

Marco Polo, of course, was a merchant, not a cleric. He had no problem questioning a learned tradition and telling its proponents that their unicorn was not a real unicorn. Other travellers, such as the fourteenth century missionary, Jordan Catala from Sévérac, in the south of France, resolved the contradiction between the animal he saw and the animal mentioned in books in the exact opposite way. After giving a rather traditional description of the rhinoceros, Jordan Catala dismisses the problem altogether: *non tamen est unicornis*, "however, it is not a unicorn" (Gadrat 2005, 249, § 36). Basically, the learned cleric is saying that *his* unicorn is not a *real* unicorn, giving priority to the bookish creature (nobody has ever sighted) over the beast he has actually faced. Again, we have a terminological issue where one term—unicorn—is said to be improperly used for one of its two "variants". (Relevant discussions were published in Einhorn (1976) on the unicorn in literature and art, Hubert (1957) and Okubo (1997) on form and size, and Ribémont (2008) on terminology.)

Challenges in the Identification of Ordinary Animals

Even if the problem of identifying and naming an animal is particularly acute when exotic or imaginary creatures are involved, the same problem occurs every time a writer does not have anything to refer to in his or her own culture and language. A domestic or semi-domestic animal can present the same kind of problem if the translator does not know which creature the text is referring to, and it is necessary to make the same leap to fill the gap as if discussing leopards, panthers, and unicorns. For the modern scholar, the challenge lies in the correct interpretation of these complex cases.

A simple example might illustrate this point. In one of the antique fables by Phaedrus, a Greek poet writing in Latin in the first century, an eagle catches a turtle but cannot get to the fleshy parts because the turtle is hiding in its shell. A malicious crow offers to tell the eagle how he can manage to eat his prey if he promises to share. The eagle accepts the terms of the contract and the crow suggests dropping the turtle from a great height down onto the rocks, where its shell would be shattered and the turtle's meat made accessible. The eagle follows the crow's advice and the two birds feed on the turtle. (For a more detailed comment and additional bibliography, see Trachsler 2002.)

The turtle, in Phaedrus' Latin text, is *testudo*. Most medieval renderings of this fable replace the term with a synonym of "shell": *Concha* in Latin, *welk* in Anglo-Norman, in French *oystre* "oyster" or *limace* "snail"—an interesting choice, because it shows that in Old French a *limace* is not necessarily without a shell as in Modern French (Möhren 1986, 202). Only the Englishman Caxton translates the term with *nutte*, "nut". The word *testudo* actually already had the meaning of "shell" in Classical Latin. However, its substitution in medieval texts was more likely triggered by the fact that the adaptors or translators of Phaedrus, almost 1000 years later and hundreds of miles from the Mediterranean world of the Greek writer, would not think of a turtle, but rather a shell. To them, the turtle was not a familiar animal, so they replaced it. It cannot be ruled out that they actually knew that Phaedrus was talking about a turtle but searched for an animal that their readers could relate to more easily. However, the scant evidence we can gather points in a different direction: The word *testudo* has left little evidence for the sense of "turtle" in the Neo-Latin languages. Vernacular terms like *tortuga* or *tartaruga* are completely unrelated to it. They are medieval creations that can be traced back to *tartarucus* "from the Tartarus", "infernal", because the turtle was somehow associated with the Other World. *Testudo*, in the Middle Ages, mostly translates as "shell" or "snail". (See Trachsler 2002, 164, notes 7 and 8, for some of the exceptions.)

Since in Northern Europe, turtles were not as common as shells or snails, they were removed from the literary tradition and, presumably, to some extent, also from the linguistic *fauna* altogether. A bilingual English-Latin glossary goes as far as to translate *Hec tortuga*, which is indeed the medieval Latin term for "turtle", with *Ae wylke*, that is, a "marine gastropod mollusc of the genus *Buccinum*, having a turbinate shell" (Wright 1884, vol. I, col. 641–62. The *Oxford English Dictionary* (1989, vol. XX, p. 206b) offers an excellent account of the history of this word in the English language).

The same kind of procedure is found almost every time a slightly technical or specific term travels over time and space. Sometime in the last third of the twelfth century, the same Phaedrian fable was translated by Marie de France into French. Marie, as far as is known, was from the western part of France and writing at the English court. The variety of French she used was Anglo-Norman. She, thus, naturally translated *testudo* as *welk*, a very common term in Anglo-Norman that was also used in Middle English and that made its way into modern English with the exact same meaning. *Welk* was not, however, easily understood in the rest of France and the copyists and illustrators of the *Fables* of Marie de France were puzzled —or felt their readers might be— when encountering the term. They, more or less, guessed what the word might mean and replaced it with continental equivalents, including *mole* "mussel", *oystre* "oyster", *miche* "shell", but also *poisson* "fish" and, again, just like Caxton, *noe* "nut". *Stricto sensu*, none of the translations is correct, since none of the shellfish mentioned is a *welk* with its "turbinate shell", but they all function within the fable. Translators used the narrative context of the fable to realize the double leap from *testudo* to "shell", and then from *welk* to "mussel", "oyster", "shell", "fish", and "nut". Similar examples and other variants may frequently be found in medieval manuscripts (Trachsler 2006). Both leaps were fairly successful because the distance separating the first item from the second is not immense. The translators could come up with an acceptable equivalent using the narrative context they had at their disposal and using an animal available in their own cultural and linguistic context that could fill in for the mysterious *testudo* and *welk*.

This is a recurrent mechanism when "translating" animal names. As long as there is a narrative context, the modern scholar can check the translation and, thus, explain that the substitution, in these cases, does not imply synonymy: in the Middle Ages, *welk* did not really mean "mussel", "oyster", "shell", "fish", and "nut", and, similarly, a rhinoceros is not necessarily a unicorn. These are just the terms that medieval authors used because they did not have anything closer at their disposal.

The situation that really challenges the modern scholar is when there is no context, such as in glossaries and wordlists, which are just naked translations of one language into another. One would expect these lists, which were used to learn Latin or, less frequently, foreign vernacular languages, to be of the greatest help when trying to determine the meaning of a given zoonym. Sometimes they are, but occasionally they contain very surprising translations that make—or should make—the modern scholar skeptical. (A perfect example was published by Evans (1993) in the form of an excellent discussion of bird names.)

In the notoriously complicated matter of fish, an Anglo-Latin glossary of the fifteenth century offers translations like *Hec delphin*: *A*^{e} *sawmone* ;

Hic polupus: *A^e lopstere*; *Hic torpedo* [= stingray], *est piscis habens multos pedes*; *Hic rumbus* [= turbot]: *A^e storjon*, and the already mentioned *Hec tortuga*: *A^e wylke* (Wright 1884, vol. I, col. 641–42). In the section on *vermium*, *Hec salamandra*, *A^e criket* may be found (Wright 1884, vol. I, col. 641), which does not seem correct, although it also occurs in the *Traitiz* by Walter of Bibbesworth (2009, v. 721). Other examples in English are provided in the *Oxford English Dictionary* (1989), while French evidence can be found in Roques (2000). The three consecutive entries for *pardus*, *Hic pardus*: *A^e leparde* ; *Hec parda*, *est femella* ; *Hic pardus*, *est qui generat cum leona* (Wright 1884, vol. I, col. 639) exemplify the confusion regarding the difference between *leopard* and *pardus*, which are first presented as synonyms and later as distinct beasts. In another manuscript, the Latin *cimex* "maggot" is glossed by the French *garance* (some kind of plant), for which Hunt (1991, vol. I, 21, note 19) presents an ingenious solution: "This puzzling gloss is explained by the fact that *cimex* ('grub, maggot') is glossed in Old English by *mada* (*made, madu*) which has been confused with *mæd(e)re* 'madder' (which translates *sandix*) which in Old French is called *garance / warance*. This also explains the gloss '*vermiculum*: garance' in T[obler]-L[ommatzsch], 4, 100". In any case, such translations are highly suspicious.

The Portrait of a Medieval Animal

One can tell that even learned clerics struggled when they did not have a context to build their translations on and also by all the blanks that some manuscripts contain. As an example, MS. London, BL, Stowe 57, which contains a list of nouns occurring in Isidore of Sevilla's *Etymologies*, has many empty spaces in the sections *De bestiis* and *De minuis animantibus*, awaiting interlinear glosses. In fact, it only contains four glosses, which are concerned with extremely common animals: *ursus*: "bere" / *canis*: "id est hund" / *catuli*: "id est whelpes" / *vulpes*: "id est fox" (Hunt 1991, vol. I, 22). None of the difficult items have been glossed.

Glossaries are, of course, of great value to lexicographers, but they need to be examined carefully and some translations should be questioned by scholars. In any case, and this should be the bottom-line of every lexicographical inquiry, the information conveyed by "errors" is just as important as a "correct" translation would be, as it reveals how a word was interpreted in a given place and time.

At this point, and before concluding, it might be interesting to test the means at our disposal when trying to draw the portrait of a medieval animal. Since fish, birds, and insects present a notoriously complex matter, I will concentrate on a non-exotic mammal, the lynx, of which one can safely assume that some form of information was available to writers in the West (Fig. 8.1). However, as shall be demonstrated shortly, outlining the silhouette of the medieval lynx almost inevitably leads one to also take into consideration the figure of its more mysterious and exotic *alter ego* –the *ounce*– and another creature, the *lupus cervarius*, that somehow overshadows them both. The lynx does not come alone. Just as the leopard needs to be examined in conjunction with the lion, the pard, and even the panther, and the unicorn demands to be seen in its relation with the rhinoceros, the lynx and the *ounce* are interwoven to the point of becoming a single animal. (For are more detailed discussion, see Trachsler 2017.)

First, it is important to remember that, as in the cases of the medieval panther or leopard, the modern ounce, an animal more commonly known as the snow leopard (*Panthera uncia*), is completely unrelated to its

Fig. 8.1 Images of lynx in the *Concordantiae caritatis* (1349–1351), Library of the Cistercian house of Lilienfeld (Lower Austria), cod. 151, fol. 31v. Photo: Institut für Realienkunde, University of Salzburg (Austria)

medieval counterpart of the same name, which is *once* or *lonce*. Its etymology does tie it to *lynx*, though, which is probably the starting point for the terminological and, hence, ontological confusion.

The Romance word *once* most probably stems from *lonce* "lynx" after apheresis, because the initial *l* was misinterpreted as the elided form of a definite article *l'*. *Lonce* is said to be derived from Italian *lonza*, "panther", adapted, during the time of the Crusades, directly from the Greek *λυνξ*, pronounced *lúnx*, with an [u], by merchants in Asia importing fur and wild cats to the West; aside from pelts, there was also a market for living animals, mostly in noblemen's courts (see *Trésor de la Langue Française informatisé*, *s.v.* "2 once" and, for additional bibliographical material, Trachsler 2017, note 14).

One immediately sees the relationship between *lynx* and *lonce*, but also the problem it must have created: if this etymology is correct, it simply reflects the dilemma of the Westerners when facing the panther. Lacking an appropriate term, they used the closest they had to describe a spotted cat, *lynx*, just as a continental scribe would use *moule* or *oistre* for *welk*. *Lynx* came close enough to the beast they were capturing or buying, but it was not, of course, a lynx in the zoological sense of the term.

However, the term *lonze* was at that time widely employed in the West to designate any sort of spotted feline, including cheetahs and hunting leopards (Buquet 2011). This point was clearly shown by Albertus Magnus, who certified that *lunza* is used not only by Romance-speaking nations, but also by Germans, as the standard term in use throughout Europe:

> *Alfech Arabice est animal, quod multi Ytalicorum et Alamannorum et Gallicorum lunzam vocant; natum autem est ex leone et leopardo et est ferox nimis et nocivum.*
>
> *Alfech* in Arabic is an animal which many Italians, Germans, and French call *lunza*; it is the off-spring of the lion and the leopard and is very ferocious and harmful. (Albertus Magnus 1920, lib. 22, tract. 2, cap. 1, 2, 1356)

Up to this point, everything seems simple: there is the Western lynx and the Eastern (l)once. But curiously, the first multilingual dictionaries of the sixteenth century translate *lynx* as *(l)once* and introduce *loup-cervier* as a third synonymous term. In 1552, to mention just one example, Charles Estienne defines the Latin word *lynx* as follows:

> *une beste du genre des loups cerviers, de veue fort ague, lequel ha diverses couleurs, et la peau tachetee de moucheture noire fort drue.*
>
> an animal of the genus of deer-wolves, with piercing eyesight, and different colours, and whose fur is marked by many black spots. (Charles Estienne 1552, *s. v.* "Lynx")

The French translation he adds is *un once, ou Lonce, ou Lunce*. So, when did the two terms, *lynx* and *lonce*, become synonymous, and are they really synonymous? If one tries to trace the evolution of these names leading up to the Renaissance, it becomes immediately clear that there is no *lonce* to be found in medieval bestiaries or encyclopedias. Only the *lynx* and the *lupus cervarius* are mentioned, and always together. Isidore, drawing on Pliny and Solinus as usual, tells us why:

> *Lincis dictus quia in luporum genere numeratur: bestia maculis terga distincta ut pardus, sed similis lupo; unde et ille λυκοσ, iste lincis.*
>
> It is named *lincis* (*lynx*) because the animal is part of the family of wolf (*lupus*): the beast has spots on its back like the *pard*, but it looks like a wolf; that is why the latter is named λυκοσ, and the former *lincis.* (Isidorus Hispalensis 1983, 102–104)

He also mentions the sharp eyesight of the *lynx/lupus cervarius* and adds that its urine turns into a precious stone with marvelous properties called *ligurius.* This stock of information remains virtually unchanged throughout the medieval encyclopedic tradition and consolidates what we find in bilingual wordlists: *Linx: une beste tres cler voant lupus ceruarius* (Roques 1936–1938, vol. I, 235). This is hard evidence of the two terms being synonymous. The most striking element is that the lynx, which definitely looks like a cat, is somehow perceived as a wolf. This is a constant feature in the iconographic tradition of bestiary manuscripts as well, where the lynx is always represented in a canine and never feline form and can only be identified because it is portrayed as urinating and producing the *ligurius* stone. One can deduct from this fact that not many clerics had actually seen a lynx and that they would just go on copying what they found in their sources, as in the case of the unicorn. The lack of direct knowledge regarding the lynx can also be seen when the twelfth-century poet Guillaume le Clerc (1892, v. 2030) translates "panther" with *love cervere* and when Philippe de Thaün (1900, v. 1178–1180) offers the same translation for "hyena". (In this context the line from the *Vita sancti*

Rainerii mentioned by Du Cange (1883–1887, t. 5, col. 141a) *s. v.* "Lonza 2:" "*In ipso deserto reperit duas hyænas, quas vulgus vocat Lonzas, leone velociores et audaciores*", ...two hyenas, commonly called Lonzas..., is very interesting.)

Any spotted animal can be called *love cervere*. The only medieval author stating that such a designation is completely misleading is a hunter, Gaston Phebus, a man who actually knew what a lynx looked like: *loup cervieres et les autres chaz loux* [...] *pourroit [on] mieulz apeler chaz lieparz que autrement*, "the animal that some call deer-wolves and others call cat-wolves should best be called leopard-cats", he complained (Gaston Phébus 1971, 104).

Interestingly in the fourteenth century, he suggests *lieparz* should be used, not *lonce*. Indeed, before the sixteenth century, the term *lonce* is very rare in France. It occurs only in the thirteenth-century *Renart* tradition, where it applies to an unidentified beast, and as a technical term in accounts of fur traders, where it seems to be used as a generic expression to designate skins of cheetahs, leopards, and other spotted felines (Delort 1978, vol. 1, 170). It becomes more widely used through Italian influence in the fourteenth century. In Italian, the word occurs quite frequently but its definition is controversial, as can be seen by the observations of the Italian and Latin commentators on Dante's *Inferno*, where the *lonza* appears in a famous sequence alongside a lion and a she-wolf in the same passage (Cassell 2010, 85–89 and 1989, 45–76). Most commentators agree that the animal in question here is a panther or a leopard. (See the Dartmouth Dante Project: https://dante.dartmouth.edu where nearly all Dante commentaries can be found.) However, when one reads that the *lonza* is the friend of all animals, it becomes clear that we are dealing with the medieval panther, and that the word actually applies to any spotted cat, which would, of course, also include the lynx. The most inclusive definition is provided by Benvenuto da Imola (1375–1380):

> *Sed ad aperiendam viam, primo, quaero quae fera sit ista lontia. Ad cujus intelligentiam vel cognitionem est subtiliter praenotandum, quod tria sunt animalia praecipue habentia pellem variis maculis distinctam, scilicet lynx, sive lynceus, qui vulgariter dicitur lupus cerverius, pardus, et panthera. Modo dico quod per lontiam autor potest intelligere lyncem, per quam figurat luxuriam.*
>
> First, one should inquire what kind of beast this *lontia* might be. In order to understand and to know this, it should be kept in mind that there are

> mainly three animals with spotted fur, *i. e.* the *lynx*, or *lynceus*, which is called deer-wolf in vernacular speech, the pard, and the panther. I suggest that in this case the author refers to the lynx, because it represents lust. (https://dante.dartmouth.edu/search_view.php?doc=137551010310&cmd=gotoresult&arg1=0)

Quite clearly, *lonza* does not refer to any specific animal for the commentators of Dante, otherwise there would be a larger consensus amongst them. *Lonza* "stands for" some other animal: the lynx, the pard, or the panther, or, in other terms, for the three felines that share the characteristic of having spotted fur. There is little more the philologist can do on this general level.

What can be done, however, is to examine the actual passages in which a word is used and the relationship of one term to others. An explanation does exist for why, at a certain moment, the French started using *(l)once* as a synonym for *lynx* and not before. But in order to get there, it is necessary to examine documents usually not taken into consideration, such as account books, laws, and sometimes cook books; we also need to consider illustrations and call on our colleagues from other departments who are more knowledgeable than we are about trade, food, clothing, iconography, but also archaeology and zoology. Only this way can scholars hope to attempt to reconstruct the identity of a medieval animal. Just opening a dictionary will not do.

Conclusions

Given the aforementioned complexities of identification, an interdisciplinary approach might be the most promising way to move forward: in order to adequately appraise the perception of animals, both humanists and natural scientists need to double-check their understanding of animals from each other's point of view. In addition to the obvious function of helping us "not get it wrong", interdisciplinarity provides a broader and thus a more accurate picture. When archaeozoologists find a bone, they must find out to which animal it belongs, and might establish that it belongs to a leopard (Bartosiewicz 2001). But that will not provide all the answers as to why and how the bone ended up in a specific site at a specific moment. A zoologist can provide background information on the habitat of the animal and establish if the animal is local or not, which might imply a historical dimension. When did leopards disappear from the European scene?

Should the animal come from afar, the historian will be expected to contribute some information on trade, and may suggest that the animal was part of attire or possibly kept in an animal park for representation. Together with the literary scholar, the historian might then be able to figure out why this specific animal, rather than another, was chosen to be incorporated into the animal collection, and so on. None of us has the entire answer, but each of us holds part of it. Understanding the perception of animals in human society is a process rather than something obtained by turning to a single source.

References

Albertus Magnus. 1920. *De animalibus libri XXVI*, edited by Hermann Stadler. Münster: Aschendorff.

Bartosiewicz, László. "A leopard (*Panthera pardus* L. 1758) Find from the Late Middle Ages in Hungary." In *Animals and Man in the Past: Essays in Honour of Dr. A.T. Clason*, edited by Hijlke Buitenhuis, and Wietske Prummel, 151–160. ARC-Publicatie 41. Groningen: Archaeological Research and Consultancy, 2001.

Buquet, Thierry. "Le guépard médiéval, ou comment reconnaître un animal sans nom." *Reinardus* 23 (2011). 12–47.

Cassell, Anthony K. "The Three Beasts." In *Lectura dantis americana. Inferno I*, 45–76. Philadelphia: University of Pennsylvania Press, 1989.

Cassell, Anthony. "The Three Beasts." In *The Dante Encyclopedia*, edited by Richard H. Lansing, 85–89. London–New York: Garland, 2010.

Delort, Robert. 1978. *Le Commerce des fourrures en Occident, à la fin du Moyen Âge (vers 1300-vers 1450).* Bibliothèque des Écoles Françaises d'Athènes et de Rome 236, 2 vols. Roma: École Française de Rome.

Du Cange, Charles du Fresne. 1883–1887. *Glossarium mediae ac infimae latinitatis*, edited by Léopold Favre. Niort: Favre. Accessed on 3.09.2018. http://ducange.enc.sorbo.nne.fr/

Einhorn, Jürgen W. 1976. *Spiritalis Unicornis. Das Einhorn als Bedeutungsträger in Literatur und Kunst des Mittelalters.* Münstersche Mittelalter-Schriften 13. München: Fink.

Estienne, Charles. 1552. *Dictionarium latino-gallicum.* Paris: apud Charles Estienne.

Evans, Dafydd H. "Taxonomy of Bird-Naming in Anglo-Norman and in Channel Island Patois." In *Anglo-Norman Anniversary Essays*, edited by Ian Short, 105–134. Anglo-Norman Text Society Occasional Publications Series 2. London: Anglo-Norman Text Society, 1993.

Faucon, Jean-Claude. "La représentation de l'animal par Marco Polo." *Médiévales* 32 (1997). 97–117.

Gadrat, Christine. 2005. *Une image de l'Orient au XIVe siècle. Les "Mirabilia descripta" de Jordan Catala de Sévérac*. Mémoires et Documents publiés par l'Ecole des Chartes 78. Paris: École des Chartes.

Gaston Phébus. 1971. *Livre de chasse avec introduction, glossaire et reproduction des 87 miniatures du manuscrit 616 de la Bibliothèque nationale de Paris*, edited by Gunnar Tilander. Cynegetica 18. Karlshamn: Johansson.

Guillaume le Clerc. 1892. *Le Bestiaire. Das Tierbuch des normannischen Dichters Guillaume le Clerc*, edited by Robert Reinsch. Altfranzösische Bibliothek 14. Leipzig: Reisland.

Hubert, Martin. "Notes de lexicographie thomiste II: La taille de la licorne." *Archivum latinitatis medii aevi* 27 (1957). 167–187.

Hunt, Tony. 1991. *Teaching and Learning Latin in Thirteenth-Century England*. Cambridge: D. S. Brewer.

Isidorus Hispalensis. 1983. *Etymologiae, Livre XII*, edited by Jacques André. Paris: Les Belles Lettres.

Levy, Brian J. "Un bestiaire oriental? Le monde animal dans *Le Devisament dou monde* de Marco Polo." In *Les animaux dans la littérature*, *Les animaux dans la littérature. Actes du Colloque de Tokyo de la Société Internationale Renardienne du 22 au 24 juillet 1996 à L'université de Keio*, edited by Hideichi Matsubara, Satoru Suzuki, Naoyuki Fukumoto, and Noburu Harano, 160–178. Tokyo: Keio University Press, 1997.

Marco Polo. 1982. *Milione. Le Divisament dou Monde. Il Milione nelle redazioni toscana e franco-italiana*, edited by Gabriella Ronchi, introduction by Cesare Segre. I Meridiani. Milan: Mondadori.

Marco Polo. 1998. *La Description du Monde*, edited/translated and annotated by Pierre-Yves Badel. Lettres Gothiques. Paris: Le Livre de Poche.

Marco Polo. 2004. *Le Devisement du monde*, general editor Philippe Ménard, t. III, edited by Jean-Claude Faucon, Danielle Quéruel, and Monique Santucci. Textes Littéraires Français 568. Genève: Droz.

Marco Polo. 2005. *Le Devisement du monde*, general editor Philippe Ménard, t. IV, edited by Joël Blanchard, and Michel Quéreuil. Textes Littéraires Français 575. Genève: Droz.

Marco Polo. 2009. *Devisement du Monde*, general editor Philippe Ménard, t. VI., edited by Dominique Boutet, Thierry Delcourt, and Danielle James-Raoul. Textes Littéraires Français 597. Genève: Droz.

Möhren, Frankwalt. 1986. *Wort- und sachgeschichtliche Untersuchungen an französischen landwirtschaftlichen Texten, 13. 14. und 18. Jahrhundert. Seneschaucie, Menagier, Encyclopédie*. Beihefte zur Zeitschrift für romanische Philologie 197. Tübingen: Max Niemeyer.

Okubo, Masami. "Encore la licorne." In *Les animaux dans la littérature. Actes du Colloque de Tokyo de la Société Internationale Renardienne du 22 au 24 juillet 1996 à L'université de Keio. Les animaux dans la littérature. Actes du Colloque de Tokyo de la Société Internationale Renardienne du 22 au 24 juillet 1996 à*

L'université de Keio, edited by Hideichi Matsubara, Satoru Suzuki, Naoyuki Fukumoto, and Noburu Harano, 255–263. Tokyo: Keio University Press, 1997.

The Oxford English Dictionary 1989, edited by John Andrew Simpson, and Edmund S. C. Weiner. Oxford: Clarendon Press.

Philippe de Thaün. 1900. *Bestiaire*, edited by Emmanuel Walberg. Lund–Paris: Möller–Welter.

Ribémont, Bernard. "La licorne, un animal exotique." *Bien dire et bien aprandre* 26, (2008). 99–119.

Roques, Mario. 1936–1938. *Recueil général des lexiques français du Moyen Âge (12e–14e siècles), I: Lexiques alphabétiques*, 2 vols. Paris: Champion, 1969–1970.

Roques, Gilles. "*Gresillon* et les dénominations du 'grillon' en français médiéval." *Travaux de Linguistique et de Philologie* 38 (2000). 7–25.

Symon Simeonis. 1960. *Itinerarium Symonis Semeonis ab Hybernia ad Terram Sanctam*, edited and translated by Mario Esposito. Scriptores Latini Hiberniae 4. Dublin: The Dublin Institute for Advanced Studies.

Trachsler, Richard. "Quelques remarques à propos du mauvais léopard dans la littérature médiévale." *Reinardus* 5 (1992). 195–208.

Trachsler, Richard. "La *tortue*, la *limace* et la *welke*. Le dossier iconographique de la fable de l'aigle et de la corneille." *Reinardus* 15 (2002). 161–174.

Trachsler, Richard. "Du *bres* au *leus*: remarques sur les variantes concernant les animaux dans les *Fables* de Marie de France." In *"Qui tant savoit d'engin et d'art". Mélanges de philologie médiévale offerts à Gabriel Bianciotto*, edited by Claudio Galderisi, and Jean Maurice, 499–508. Civilisation Médiévale XVI. Poitiers: Université de Poitiers-Centre d'Études Supérieures de Civilisation Médiévale, 2006.

Trachsler, Richard. "Du lynx à l'once. Animaux réels et créatures symboliques." *Reinardus* 29 (2017). 142–163.

Trésor de la langue Française informatisé. ATILF–CNRS & Université de Lorraine. Accessed on 27.08.2018. http://www.atilf.fr/tlfi

Van den Abeele, Baudouin. "Les stratégies olfactives chez les animaux et leur moralisation." In *Parfums et odeurs au Moyen Age. Science, usage, symboles*, edited by Agostino Paravicini Bagliani, 429–444. Collection Micrologus' Library 67. Firenze: Edizioni del Galluzzo, 2015.

Walter de Bibbesworth, 2009. *Le tretiz*. edited by William Rothwell from MS. G (Cambridge University Library Gg.1.1) and MS. T (Trinity College, Cambridge 0.2.21) together with two Anglo-French poems in praise of women (British Library, MS. Additional 46919). Texts and Publications. Aberyswyth–Swansea: The Anglo-Norman On-Line Hub.

Wright, Thomas. 1884. *Anglo-Saxon and Old English Vocabularies*, edited and collated by Richard Paul Wülcker (2nd ed.). London: Trübner.

CHAPTER 9

Exotic Encounters: Vikings and Faraway Species in Motion

Csete Katona

Introduction

People regarded as exotic from a certain viewpoint are often also associated with an exotic fauna. Such is the case with those Northmen—often labeled Vikings—who inhabited Scandinavia from the late ninth to mid-eleventh centuries, a place that could have certainly offered 'exotic' animal species to peoples in Europe at the time. In the Viking Age, large numbers of Scandinavian men and women traveled abroad and reached faraway lands outside Scandinavia, such as European Russia, the Near-East, the North Atlantic islands, and even America, which all presented the opportunity to run across a local fauna, that was unknown in the Scandinavian homelands. Bearing this twofold meaning of 'exoticism' in mind, a list of animals that can be considered 'exotic' will be provided here, containing animals which, on the one hand, would have been unusual for the contemporaries of the Vikings or, on the other hand, were encountered by Scandinavians during the course of the Viking Age (c. 800–1050) outside their homelands.

C. Katona (✉)
Department of Medieval Studies, Central European University, Vienna, Austria

L. Bartosiewicz, A. M. Choyke (eds.), *Medieval Animals on the Move*, https://doi.org/10.1007/978-3-030-63888-7_9

I have gathered information from various types of written sources (including Old Norse sagas, Muslim travel narratives and Byzantine writings) which explicitly mention Scandinavians encountering exotic species as well as a few archaeological assemblages in the Viking World that include non-local, exotic animal remains. However, as noted, what should be regarded as exotic is a matter of debate, since from a (medieval) Scandinavian point of view, not all animals presented here could be labeled as such. Whales encountered by the Norwegian settlers on the shores of Iceland, for instance, would not have been new to them, even though from an insular European perspective, they would have been regarded as exotic.

Rare animals which were not native in the North of that period, will obviously be listed, although due to the regional differences within Scandinavia, other animals can also fall into this category; for example, we can be quite sure that the Icelandic fauna would not have been equally normal for a Norwegian (who came from a similar landscape) and a Danish Viking whose original living place was the flat Jutland peninsula. Animals that were unknown in insular Europe but appeared there in some form through Viking intermediaries will also be mentioned. By its very nature, this essay is rather descriptive than analytical, and the list provided here is not full; future contributions to it will be most welcome.

Scandinavia

Domestic and wild animals were essential parts of human life in the Middle Ages in both a practical and symbolic (or religious) sense. Viking Age Scandinavia was no exception in this regard. A wide range of domestic animals were kept which were also transported abroad during the Viking voyages (Fig. 9.1). They included cattle, pigs, goats, sheep, horses, cats, and dogs. Wild animals, such as reindeer in Norway, were hunted for meat and fur or for other luxury products such as the antler used to produce hair combs (Ashby et al. 2015). Some wild animals, however, also filled an ideological role: wolves and bears for instance, were often subjects of religious beliefs and initiation rituals (Schjødt 2006; Gunnell 1995).

Scandinavian gods were also often associated with animal companions: the chief-god Odin was aided by his eight-legged horse Sleipnir, the chariot of the thunder god, Thor, was pulled by goats, and Freyja, the goddess of love and fertility, was accompanied by a cat. In mythological accounts, the two chief antagonists of the gods were animals: the wolf, Fenrir, and the snake called the Midgard-serpent. Besides other wild animals such as deer, boar, and a single squirrel, birds are also known from the *Eddas*:

Fig. 9.1 Horses unloaded from a longship on the Bayeux tapestry (Source: https://commons.wikimedia.org/wiki/File:BayeuxTapestryScene39.jpg [accessed: 01.02.2018.])

ravens acted as scouts of Odin, while Ragnarök, the end of the world, was signaled by the crow of a rooster (von Hofsten 1957). Apart from the animals enumerated above, however, Vikings had a much wider knowledge of animal species due to their extensive voyages outside mainland Scandinavia.

Exotic Animals from the 'West'

When the Viking expansion began westward in the ninth century, the first place where an 'exotic' fauna could be encountered was Iceland, and then later, Greenland. Prior to the arrival of the Northmen, the only land mammal found in Iceland was the arctic fox, which most likely reached the island from the Arctic on detached sea ice. The island's birdlife, on the other hand, was enormously rich. Several exotic birds could also fly there from distant locations. Of these, however, only swans, falcons, sea eagles, and ptarmigans stayed for the winter (Smith 1995, 323). While some of these birds might have been known by the Norwegian or Celtic settlers,

Scandinavians from other areas would certainly have found them exotic. The island's special geography also presented new challenges for the settlers and, thus, hunting practices for these animals might have been different compared to those pursued in their homelands. The ptarmigan for instance, was usually hunted using nets according to the late medieval Icelandic evidence found in *Droplaugarsona saga* (Jóhannesson ed. 1950, 145) and *Fljótsdæla saga* (Jóhannesson ed. 1950, 242–245), but new methods also developed. According to the latter source, some people 'did not hunt ptarmigan the same way as other men. They had no nets and shot [them] with javelins' (translation by the author, Jóhannesson ed. 1950, 242).

Similarly, whales appeared in a new economic and social context in Iceland. This mighty sea mammal possessed critical resources in the huge amount of fat and meat it provided. Since carcasses of whales were often washed ashore on the island's coast, they were not hunted. This practice seems to differ from what the Norwegian traveler, Ohthere, recorded in the ninth century about Norway: 'But in his [Ohthere's] own country is the best whaling; those [whales] are forty-eight ells long, the largest, fifty. He said that with five companions, he had slain sixty of these in two days' (Cross 1981, 20–21). Whales also became a major source of conflict in Iceland between nearby farmers who even took up arms against each other over their valuable carcasses (Byock 1982, 229–231). Icelandic sagas, produced 200 or 300 years after the events they describe, are not always to be trusted as authentic accounts for the Viking Age, although these traditional ways of hunting might have changed little in the island since then. Unusual circumstances and methods of dealing with animals, old or new alike, such as noted with ptarmigans and whales, could have easily evolved in new environments.

Not only human reactions toward known species changed in a new environment, but animals transported to distant lands also gained new significance. Even though also present in Iceland during the early years of the settlement (McGovern 2011, 5; Fig. 9.2) prior to its extinction there, the walrus became the first and foremost exotic animal for settlers of the Greenland colonies (Frei et al. 2015). Luxury products made from walrus ivory spread widely into early medieval Europe through the mediation of Scandinavian traders (e.g. the Lewis chessmen in Britain; Robinson 2004). However, this valuable raw material must have been known in Norway from the early Viking Age. According to Ohthere's account, he presented

Fig. 9.2 Walrus tusk from Iceland (McGovern 2011; With the author's kind permission)

King Alfred of Wessex with a piece taken from the Baltic (Cross 1981, 19–21). Nevertheless, walrus ivory represented an exotic product in Europe, and how far it spread from the traditional trading routes during the Viking Age is illustrated by the hilt of the famous Viking sword of St. Stephen I, King of Hungary. The sword in question is kept in the St. Vitus cathedral of Prague and, according to previous (quite early) evaluations which were repeated by later research, its hilt was made of elephant ivory (László 1977; Fodor 2003–2004). However, it has never been properly examined using modern equipment (as far as I am aware) and the sword is a typical Viking one, decorated in the Mammen-style and thus, the use of walrus ivory is much probable. This is even more so since a crosier from the Veszprém convent in Hungary also proved to be made of walrus ivory after proper zoological examination (Fig. 9.3; Fülöp and Koppány 2004; Choyke 2013, 3). In the twelfth century, precious walrus tusks were even exported as far as Iran and India (Keller 2010, 8).

Another exotic product from Greenland was the polar bear, which according to *Landnámabók*, a genealogical work of the settlers of Iceland, was unknown in Norway and could have reached Iceland similarly to the arctic fox, namely by floating on ice (Benediktsson ed. 1986, 219). The polar bear was used in its 'living form' and was presented live as a prestige gift to rulers of courts within Scandinavia according to Icelandic sagas—*Auðunar þáttr vestfirzka* (Þórólfsson and Jónsson eds. 1943), *Vatnsdæla saga* (Sveinsson ed. 1939, 44), and *Króka-Refs saga* (Halldórsson ed. 1959, 142–143). They were highly appreciated even outside Scandinavia: Ísleif Gizurarson, the first bishop of Iceland, gifted the Holy Roman Emperor Henry III with a polar bear while on pilgrimage (Egilsdóttir ed. 2002, 7).

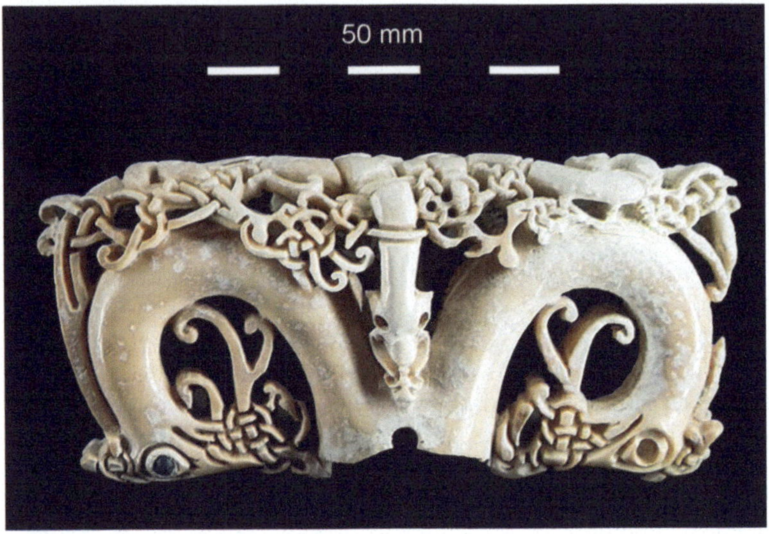

Fig. 9.3 Walrus tusk crosier from the Veszprémvölgy convent in Hungary (Fülöp and Koppány 2004. With the authors' kind permission)

Exotic Animals from the 'East'

The *Austrvegr*, that is, 'the Eastern Way'—as Scandinavians called the territories from Eastern Europe through Byzantium to the Near-East—certainly possessed a more exotic fauna than what Vikings could have found in the West. In European Russia, the Vikings used the waterways for transport, and these rivers as well as their surrounding habitats presented new animal species to someone coming from the North.

The first group of these animals which should be listed are the small carnivores and rodents which were widespread in the Baltic and the Northern Russian territories. The fur from these animals was the main product Scandinavians exchanged for the Islamic *dirham* coins in markets such as Bulghar or Itil along the River Volga. According to the thirteenth-century *Egils saga Skallagrímsonar* (Nordal 1933, 27), furs were taken by force or collected in the form of taxes in Norway from the Lapps as well (Nordal ed. 1933, 34–42). However, additional new species must have been present in the 'East'. The Arabic writer al-Maqdisi (ca 985) listed sable, grey squirrel, ermine, mink, fox, marten, beaver, and spotted hare among animals whose skin was traded in Bulghar (Lunde and Stone 2012,

169–170). This list can be amended by another Muslim author's statement, namely al-Masudi's. According to his account, the pelt of the black fox was the most valuable (Lunde and Stone 2012, 136).

Before the markets of Bulghar and Itil became established in the trade system of the 'East', there is evidence that Vikings ventured as far as Baghdad with their wares on camel back (Pritsak 1970), an animal that would certainly have been an exotic mount for them.

Even though snakes were known in Scandinavia, during their passage on the Volga, Vikings might have encountered local species of snakes which were dwelling in great numbers along the river, according to the tenth-century Arabic diplomat Ibn Fadlan (Lunde and Stone 2012, 33–34).

Even more notable is an animal from the Volga area which even surprised Ibn Fadlan himself. The vivid description of the Arabic traveler clearly illustrates the surprise a Viking traveler might have experienced upon encountering an 'exotic' animal:

> an animal, in size smaller than a camel, but higher than a bull. Its head, is a head of a sheep, and its tail is a tail of a bull, its body is a body of a mule, its hoofs are like the hoofs of a bull. In the middle of the head it has one thick round horn, as it raises it becomes thinner, until becomes like a tip of a spear. And some of them are five elbows to three elbows long, according to the bigger or smaller size (of the animal). It grazes on the leaves of trees with excellent sprouts. When it sees a horseman it goes to him, and if a racehorse is under him (horseman), it (the racehorse) runs for rescue from it in a fast flight, and if it would catch up with him (horseman) it grasps him with its horn from the back of the horse, then throws him into the air and catches him on its horn, and does not stop doing it until it kills him. And it causes no harm to the horses in any way. And they (the inhabitants) search for it in the steppe and forests until they kill it. So, they climb high trees between which it is. A few shooters with poisoned arrows gather for this purpose, and when it appears between them they shoot at it, until they wound it and kill it. And really, at the malik I saw three big bowls, similar to the Yemen (shells) "jaz", he told me that they (bowls) are made from the base of the horn of this animal. (Lunde and Stone, 2012, 42–43)

Ibn Fadlan himself was accused by researchers of not personally seeing Volga Bulgarian rhinos; however, local tales, like the one quoted above, not only served to attract the attention of foreign statesmen but had to reflect reality if the story was checked. Ibn Khordadbeh, another Arabic informant, already reports in the ninth century that rhinos roamed the steppe (for discussion see: Montgomery 2004).

Thus, although not explicitly associated with the Vikings, the description of the rhino by Ibn Fadlan comes from an area which was frequently visited by Scandinavians as well, raising the very great possibility that Vikings also met this animal.

If the semi-legendary saga of Yngvar the Far-Traveler is to any degree to be believed, a Viking expedition in the Volga region experienced a similar shock when they ran into an elephant with a tower on its back, led by ten local men. According to the story, the Vikings run ashore to satisfy their curiosity and inspect the unknown beast but were unable to control it until seeing how the natives direct it with reins and a cross-beam (functioning as a steering-wheel) placed in the wooden tower (Olson ed. 1912, 37–39). Although the story has an anachronistic air about it, exotic encounters with rhinos or elephants were certainly a possibility for East-farers.

The other major river route, the Dnieper, also contained exotic species. Scandinavian merchants traveling to Constantinople from Kiev and other settlements of the Rus, had to pass through seven waterfalls on the Dnieper, one of which was named *Aeifor* in Old Norse and *Neasit* in Slavonic because pelicans nested in the stones of the barrage (Moravcsik 1967, 58–59).

In addition, Vikings would not only have become acquainted with local species along the *Austvegr*, since animals from even more distant places reached territories travelled by Scandinavians. The statue of the famous Pireus lion (now in Venice) for instance is engraved with Nordic runes, testifying to the wide-ranging knowledge of exotic species that could be acquired by Scandinavians serving in Byzantium (Snædal 2014). Other exotica might be reflected by a monkey skull discovered in Rjurikovo Gorodische (Brisbane et al. 2007), a town in Rus' inhabited by Northmen. A peacock which was found in one of the most precious Scandinavian ship burials, Gokstad (Norway), also deserves mention. The presence of this rare animal most likely signaled the prestige of the buried ruler, demonstrating his access to exotic products (Dobat 2015). The peacock, originally coming from India, may even have reached Iceland, or at least might have been known there by some individuals, judging by a nickname appearing in the Icelandic *Laxdœla saga*, in which one of the protagonists, Olaf Höskuldsson, was labeled *pái*, a 'peacock', because of his lavish dressing style (Sveinsson ed. 1934, 39).

Concluding Remarks

The list presented here illustrates the wide spectrum of 'exotic' faraway species that might have been observed by Viking Age Scandinavians or about which they could spread knowledge during their voyages. This repertoire demonstrates a broad range of knowledge; however, it has to be borne in mind that even though these exotic species were all associated with 'Vikings', individual experience of such animals would have varied greatly. Considering the Viking World through these exotic animals serves to emphasize the differences between one Viking and another since one's individual knowledge of animal species was restricted by the geographical areas they came from and the ones they visited.

Acknowledgments This paper was initiated within the framework of a seminar, entitled *Animal Entanglements in the Middle Ages: From Meat to Metaphor*, taught in the Medieval Studies MA-Programme of Central European University (Budapest) in 2017 by Alice M. Choyke and Gerhard Jaritz, to whom I am most thankful for their assistance.

References

Primary Sources

Benediktsson, Jakob ed. 1986. *Íslendingabók-Landnámabók*. Íslenzk fornrit vol. I. Reykjavík: Hið íslenzka fornritafélag.

Cross, Alan S. C. transl. 1981. *The Terfinnas and Beormas of Ohthere*. London: Viking Society for Northern Research 1981.

Egilsdóttir, Ásdís ed. "Hungrvaka." In *Biskupa sögur*. Vol. 2., 1–43. Íslenzk fornrit vol. XVI. Reykjavík: Hið íslenzka fornritafélag, 2002.

Halldórsson, Jóhannes. "Króka-Refs saga." In *Kjalnesinga saga*, 117–160. Íslenzk fornrit vol. XIV. Reykjavík: Hið íslenzka fornritafélag, 1959.

Jóhannesson, Jón ed. 1950. *Fljótsdæla saga*. Íslenzk fornrit vol. XI. Reykjavík: Hið íslenzka fornritafélag.

Lunde, Paul, and Caroline Stone transl. 2012. *Ibn Fadlan and the Land of the Darkness: Arabic Travellers in the far North*. London: Penguin.

Moravcsik, Gyula ed. "Constantine Porphyrogenitus." In *De administrando imperio*, translated by Romilly J. H. Jenkins. Washington DC: Dumbarton Oaks, 1967.

Nordal, Sigurður ed. 1933. *Egils saga Skallagrísmsonar*. Íslenzk fornrit vol. II. Reykjavík: Hið íslenzka fornritafélag.

Olson, Emil ed. 1912. *Yngvars saga víðförla. Jämte ett bihang om Ingvarsinskrifterna*. København: S. L. Møllers Bogtrykkeri.

Sveinsson, Einar Ólafur ed. 1934. *Laxdœla saga*. Íslenzk fornrit vol. V. Reykjavík: Hið íslenzka fornritafélag.

Sveinsson, Einar Ólafur ed. 1939. *Vatnsdæla saga*. Íslenzk fornrit vol. VIII. Reykjavík: Hið íslenzka fornritafélag.

Þórólfsson, Björn K. and Guðni Jónsson eds. "Auðunar þáttr vestfirzka." In *Vestfirðinga sögur*, 361–368. Íslenzk fornrit vol. VI. Reykjavík: Hið íslenzka fornritafélag, 1943.

Secondary Sources

Ashby, Steven P., Ashley N. Coutu, and Søren M. Sindbæk. "Urban Networks and Arctic Outlands: Craft Specialists and Reindeer Antler in Viking Towns." *European Journal of Archaeology* 18, no. 4 (2015). 679–704.

Brisbane, Mark, Ellen Hambleton, Mark Maltby, and Evgenji Nosov. "A monkey's tale: the skull of a macaque at Ryurik Gorodische during excavations in 2003." *Medieval Archaeology* 51 (2007). 185–191.

Byock, Jesse L. 1982. *Feud in the Icelandic Saga*. Berkeley–Los Angeles–London: University of California Press.

Choyke, A. M. 2013. "Hidden Agendas: Ancient Raw Material Choice for Worked Osseous Objects in Central Europe and Beyond". In *From These Bare Bones. Raw materials and the study of worked osseous materials*, edited by A. M. Choyke, and S. O'Connor, 1–13. Oxbow Books: Oxford.

Dobat, Andres S. "Viking stranger-kings: the foreign as a source of power in Viking Age Scandinavia, or, why there was a peacock in the Gokstad ship burial?" *Early Medieval Europe* 23, no. 2 (2015). 161–201.

Fodor, István. "Das Schwert Stephan des Heiliges." *Folia Archaeologica* 51 (2003–2004). 159–172.

Frei, Karin M., Ashley N. Coutu, Konrad Smiarowski, Ramona Harrison, Christian K. Madsen, Jette Arneborg, Robert Frei, Gardar Guðmundsson, Søren M. Sindbæk, James Woollett, Steven Hartman, Megan Hicks, and Thomas H. McGovern. "Was it for walrus? Viking Age settlement and medieval walrus ivory trade in Iceland and Greenland." *World Archaeology* 47, no. 3 (2015). 439–466.

Fülöp, András, and András Koppány. "A crosier from the territory of the Veszprémvölgy convent." *Acta Archaeologica Academiae Scientiarum Hungaricae* 55 (2004). 115–135.

Gunnell, Terry. 1995. *The Origins of Drama in Scandinavia*. Cambridge: D. S. Brewer.

Keller, Christian. "Furs, Fish and Ivory: Medieval Norsemen at the Arctic Fringe." *Journal of the North Atlantic* 3, no. 1 (2010). 1–23.

von Hofsten, Nils. 1957. *Eddadikternas djur och växter*. Skrifter utgivna av Kungl. Gustav Adolfs Akademien 30. Uppsala: Almqvist & Wiksells Boktryckeri.

László, Gyula. "Beiträge zum Schwert des hl. Stephan." In *Archäologie als Geschichtswissenschaft. Studien und Untersuchungen. (Festschrift Karl-Heinz Otto)*, edited by Joachim Hermann, 467–476. Berlin: Akademie-Verlag, 1977.

McGovern, Thomas. 2011. *Walrus Tusks & Bone From Aðalstræti 14–18, Reykjavík Iceland*. CUNY Northern Science & Education Center NORSEC Report. New York NY: City University of New York.

Montgomery, James E. "Travelling Autopsies: Ibn Fadlan and the Bulghar." *Middle Eastern Literatures* 7, no. 1 (2004). 3–32.

Pritsak, Omeljan. "An Arabic Text on the Trade Route of the Corporation of ar-Rūs in the Second Half of the Ninth Century." *Folia Orientalia* 12 (1970). 241–259.

Robinson, James. 2004. *The Lewis Chessmen*. Objects in Focus Series. London: British Museum.

Schjødt, Jens Peter. "The Notion of *Berserkir* and the Relation between Ódinn and Animal Warriors." In *The Fantastic in Old Norse/Icelandic Literature. Sagas and the British Isles: Preprint Papers of the Thirteenth International Saga Conference. Durham and York 6th–12th August, 2006*. Vol. 2., edited by John McKinnell, David Ashurst, and Donata Kick, 886–892. Durham: The Centre for Medieval and Renaissance Studies, 2006.

Smith, Kevin P. "Landnám: The Settlement of Iceland in Archaeological and Historical Perspective." *World Archaeology* 26, no. 3 (1995). 319–347.

Snædal, Thorgunn. 2014. *Runinskrifterna på Pireuslejonet i Venedig*. Riksantikvarieämbetet, Stockholm.

CHAPTER 10

The Question of Feathers in the Early Modern Cabinet of Curiosities (Sixteenth–Seventeenth Centuries)

Myriam Marrache-Gouraud

Introduction

Early Modern collectors used to gather unusual objects in their private museums, which were called cabinets of curiosities. These were small rooms or chambers (or sometimes even closets) containing various wonders from the natural and human worlds. The cabinets would be full from top to bottom with items considered to be prodigies or rarities, as can be seen in the well-known engraving by Ferrante Imperato depicting his own cabinet of curiosities displayed at the Palazzo Gravina in Naples (Fig. 10.1).

Early cabinets of curiosities were designed to represent the world in the form of microcosms (as opposed to a macrocosmic representation). These collectors' attempts to encompass what they believed to be God's Creation was meant to both glorify and celebrate the Divine and interpret the enigmas of nature. The cabinets would host many a debate on the strange

M. Marrache-Gouraud (✉)
Laboratoire Forellis, Université de Poitiers, UFR Lettres et Langues, Département Lettres, Poitiers Cedex 9, France

L. Bartosiewicz, A. M. Choyke (eds.), *Medieval Animals on the Move*, https://doi.org/10.1007/978-3-030-63888-7_10

Fig. 10.1 Ferrante Imperato's cabinet of curiosities displayed at the Palazzo Gravina, Naples. From *Dell'Historia Naturale* (Imperato 1599; Wikimedia commons. Image in the public domain). https://commons.wikimedia.org/wiki/File:RitrattoMuseoFerranteImperato.jpg

manifestations of nature. Where could unicorns be found? Were they terrestrial animals or sea creatures? What therapeutic use could be made of their horns? Did chameleons only feed on the wind? In the context of this idea of wonder and of wondering about nature, feathers were of great interest to collectors because the discovery of the New World had brought not only newly discovered natural specimens with bright colors never seen before but also feathered artifacts used in the material culture of Native Americans from North, Central, and South America.

This paper aims to study how these feathers were "on the move", still attached to the animals when they arrived in Europe, in relation to the technical issues they raised in the mind of collectors. I will begin by considering the material objects concerned, examining how they became curiosities and how they arrived in Early Modern European collections. I will then show how the different display options were meant to present different answers to the numerous questions at the time surrounding classification and the organization of knowledge.

Feathers in Motion

What did collectors choose to retain from the macrocosm to compose their microcosms and reproduce the whole world in one room? As part of the natural world, feathers were an obvious choice. So, what kinds of feather items were displayed in cabinets of curiosities?

Feathers first entered cabinets in the form either of stuffed birds or as raw materials (i.e. feather and skin, with no taxidermy involved). There was a particular interest in foreign species, especially in previously unknown birds from the New World, such as the species called *Manucodiata* (or "Bird of Paradise") at the time. This species was initially brought back from the Molucca Islands, where it was thought to feed on the wind because collectors believed it had no legs. Legend had it that it ate, slept, and reproduced in flight. This apodal bird was a very curious specimen indeed. While it was later discovered that the birds' feet (a perishable part of the animal) had been cut off by mariners when preparing their dried skins so that they would not decay during the long voyage back to Europe (Swadling 1997; Schnapper 1988, 80–82), they nevertheless continued to fascinate European collectors because of the rarity of their bright colors and their curious oblong shape. Hence, this is why we find so many representations of these birds in cabinets of curiosities as well as on the garments and hats of the nobility (Swan 2015). They entered Early Modern European society through both these channels. Every collection from England to Flanders, Germany, Italy, and beyond, whether it belonged to a monarch or an ordinary citizen, would try to include at least one bird of paradise.

The toucan, well known for its immense, heavy beak and colorful feathers, was also a popular species among collectors as was the scarlet macaw parrot with its bright red feathers. Both were put on display in cabinets to amaze those who came to see them. André Thevet, King Henri II's cosmographer, came back from Brazil in 1557 with a large number of scarlet macaw parrot feathers, which he displayed with great pride in his cabinet. He deeply regretted their loss when they were unfortunately eaten away by worms and moths. The poor conditions of preservation at the time ultimately exposed the birds' skins to a great risk of decay. Their fragility (in other words the risk of losing or having them damaged) made them all the more priceless.

Feathers could also be incorporated into works of art when they arrived as raw materials. André Thevet brought a gigantic, red, feathered cape, which had belonged to an important chief, plus weapons adorned with feathers. He presented them all in tribute to the King of France, Henri II. These objects could be used as a means of distinguishing oneself and finding favor with a powerful protector. They were usually exchanged as precious gifts in Early Modern Europe and were particularly prized in Rome, France, the Germanic countries, and at the Medici court. Feather canvases were also very rare pieces. They usually came from Mexico, where the art of feathering was very elaborate. This foreign technique combined with European subjects and subject matter gave rise to hybrid objects. These are extremely interesting in that they are material testimonies of crossed cultures, hints of the first encounters and clues to post-colonial productions (Russo 2008). The transfer of curious objects from one side of the world to the other intensified (Martin and Bleichmar 2015).

Feather paintings with Christian religious subjects were commissioned and offered as gifts to successive Popes, as we can see from the example of the *Mass of Saint Gregory*, which was sent to Paul III in 1539. Very few of these feather paintings remain today. Aside from these artifacts and works of art, the two largest categories of items incorporating feathers were clothing and ceremonial objects, including crowns, necklaces, leg ornaments, weapons, and ritual or musical instruments (for instance, maracas).

We know that feathers were collected either as *naturalia* or *artificialia*. Hence, they were always on the move. For example, there was topographical movement of the animal pieces, from the exterior space that is nature, where they lived, to the interior, confined space of the private chamber, where the stuffed dead bodies would then lay on display under the gaze of the keeper and their visitors. There was also evident geographical movement, when works of art would be transferred to Europe from other continents. Moreover, there was a second kind of movement. This latter was not a matter of space but of production, namely the transformation from a natural, raw material into a work of art. Finally, the status of the piece of animal changed. Whatever they were before entering the collection, they stood as curiosities once they were included.

Their withdrawal from the natural or economic world turned them into motionless objects, "*semiophores*" removed from circulation and kept in a collection which gave them a renewed meaning (Pomian 1987). Their final movement led to stillness, fixed in place on a shelf in a museum.

The collected object could always, of course, be moved within the room whenever a display was reorganized. It could also be discarded or even sold and then displayed in another collection elsewhere according to new display options, where it might then be subject to new movements within the small space of this new collection. We will now consider these other kinds of movements inside the museums, the changing display possibilities inside the collectors' worlds.

Display Options and the Organization of Knowledge

Different kinds of objects, including feathers, implied different display options. I will now show that these display options betrayed different kinds of ideology, based on different ways of looking at both nature and art.

In the first case presented previously, when the remains of birds (i.e. their skin and feathers) were stuffed, they were displayed so as to resemble the living animal. Whether they were presented on a branch, against a fake landscape or in a flying or standing position, they imitated life. They provided a kind of illusion of nature, as if they might start to move any moment, as if they had been suspended in time in their living environment. When the bird was not stuffed, the flat skin could be kept in a special box, which was customized for each bird, as can be seen for instance in the special boxes engineered for the bird of paradise in the Queen's cabinet at Windsor Castle. This process reveals the extreme care taken of and the enormous value attributed to such pieces in a collection. These kinds of boxes were carefully arranged in drawers, coffins, or closets. In both these cases, the pieces were kept as birds and positioned alongside the other flying animals "of the air" among the three traditional categories used to organize the natural world according to the place they live in (the other categories included terrestrial animals and animals of the sea).

When the feathers had been used to make exotic objects, where was the collector to put them? The biggest items, such as the capes made out of scarlet macaw feathers, generally seem to have been hung on the walls among other spectacular items, arranged there specifically because they were so huge and magnificent that they deserved to be highlighted by the general setting.

When a collector owned various items of apparel, they could fashion a kind of "theatre" of the New World, gathering together all their *exotica* in one part of the cabinet. The feather pieces would then be shown off to their best advantage among other objects such as baskets and shoes. In these cases, the feathers took their place within a special setting that was organized from a geographical point of view. Assuming that "American Indian" meant all of America, feathers would be hung or lain alongside a hammock, canoe, or weapons or accompanying fruits, seeds, and flowers gathered from the New World. Feathers in such displays represented a "strange", other life, where they were used as clothing, something totally unbelievable from the contemporaneous European point of view (Marrache-Gouraud 2016).

Alternatively, the collector might recognize the sewing of feathers with tree barks as fine artistry and choose to associate the artifacts with other objects requiring real expertise, called *artificialia*. The feather items would then be stored alongside mathematical, mechanical, or optical and even musical instruments. This was the option chosen by Pierre Trichet in his cabinet in Bordeaux (Trichet 1631–1635).

Moreover, as there were no hard and fast rules of presentation, each collector devised their own system for classifying and arranging their collections. The final two attempts I will discuss are very interesting in that they reveal new approaches in this regard. First, feather items could be displayed alongside the birds they came from in order to show, somehow, the whole process from nature through to the making of the objects. Such displays can be found in Antonio Giganti's studio in Bologna and in his friend Ulisse Aldrovandi's treatises on birds (e.g. Aldrovandi 1600, 1610). Aldrovandi chose to also include in his discussions, all objects that had been made using parts of birds. Feathered artifacts were given special treatment in his treatise, displayed on the bodies of Native Americans, so that the readers could see how exotic these people looked when wearing such clothes (Fig. 10.2). Early modern age scholars in Europe frequently assimilated various sources to draft what were aimed to be "realistic" accounts of the New World, showing an honest effort to reliably document such curiosities. With these plates, the study of birds was itself being converted into a multidisciplinary treatise on people and their costumes. We can even see the differences between men's and women's attire.

The final attempt at classifying and organizing a collection I want to discuss came from the Roman collector Tomaso dei Cavalieri, a friend of Michelangelo. He had the idea of grouping feathered clothes and weapons with antique marble statues and religious paintings. This is very interesting because it means that he saw them both as being part of the same universal culture. According to this display option, there was no real difference between Greco-Roman ancient culture and this special genre in New World art. By joining the ancient and the new worlds together, Cavalieri was declaring that no one was barbarous. He was implying that these so-called "savages" shared a culture of their own, which was so

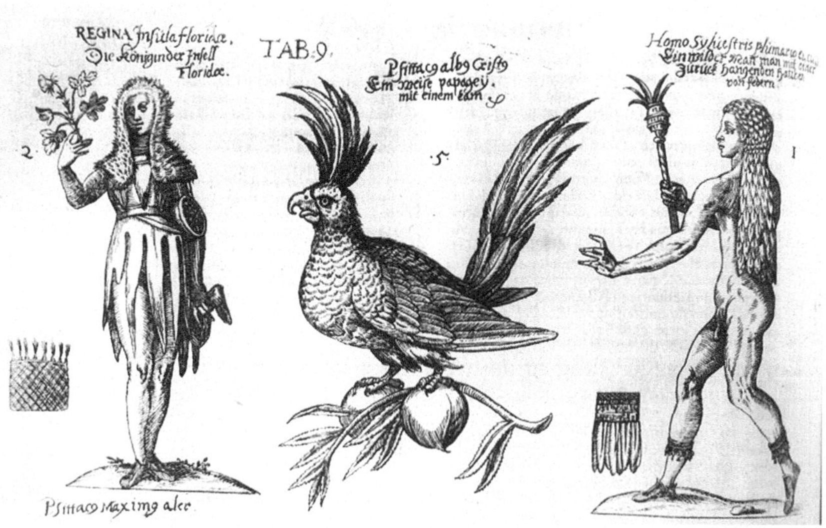

Fig. 10.2 Feathered decoration adorning the bodies of Native Americans was treated by Aldrovandi along with exotic birds of the New World (Aldrovandi 1610. L. XI, chap. 1, p. 334; ETH-Bibliothek, Zürich. Image in the public domain). https://www.e-rara.ch/zut/wihibe/content/zoom/4149659

well-developed that it deserved to be placed alongside the other prestigious humanist antiquities highly appreciated in Europe. The organization of a collection was also therefore a matter of ideology.

Conclusions

While collectors may have agreed on the relevance of collecting feathers in their different representations, there was no consensus on how best to display these items. Hence, feathers, unlike other animal parts such as bones and beaks, were curiosities that deserved a characterization "on the move". From the perspective of display options, we can conclude with the idea of a moving definition that was dependent on the collector's ideological point of view.

Aside from being regarded as a state of nature, as geographically centered or as highlighting concerns about faraway lands, savage uses, madness and cruelty, feathers could also signify an incredibly fine culture capable of creating wonders by working on unexpected materials. In this respect, the feathered artifacts could be afforded the same prestige as antiquarian objects. Should they be considered ancient or modern? On this basis, feathers stand as exemplary in the Early Modern inquiry about exoticism (Gliozzi 2000; Hartog 2005; Mason 1998; Schmidt 2015). Could they be part of our most prestigious culture? Were they just strange, exposing us to the scary unfamiliar, and completely opposed to our uses, or were they, on the contrary, extremely beautiful and therefore desirable and priceless? Feathers were not gold, emeralds, diamonds, marble or ivory. They were a costless and very fragile material, which nevertheless, had a fascinating brightness and the sumptuous rarity of precious gem stones. This ambiguity conveyed a certain idea of the beautiful savage, of art despite savagery and of nature as an artist.

References

Aldrovandi, Vlissis. 1600. *Ornithologiae tomus alter...cum indice copiosissimo variarum linguarum*. Bononiae: apud Ioannem Baptistam Bellagambam.

Aldrovandi, Vlisse. 1610. *Ornithologiae, hoc est De Avibus Historiae Libri XII*. Francofurti: typis Wolffgangi Richteri, sumptibus heredum Nicolai Bassaei.

Gliozzi, Giuliano. 2000. *Adam et le Nouveau Monde. La naissance de l'anthropologie comme idéologie coloniale: des généalogies bibliques aux théories raciales (1500–1700)*, translated by Arlette Estève, and Pascal Gabellone. Paris: Théétète éditions.

Hartog, François. 2005. *Anciens, modernes, sauvages*. Paris: Galaade éditions, Points Seuil.

Imperato, Ferrante. 1599. *Dell'historia naturale*. Napoli: Stamparia à Porta Reale, per Costantino Vitale.

Marrache-Gouraud, Myriam. "La plume, en son histoire allégorique." In *S'exprimer autrement: poétique et enjeux de l'allégorie à l'Âge Classique*, edited by Marie-Christine Pioffet, and Anne-Élisabeth Spica, 253–270. Tübingen: Narr Verlag, 2016.

Martin, Meredith, and Daniela Bleichmar. "Objects in Motion in the Early Modern World." *Art History* 18 vol. 34 (2015). 604–619.

Mason, Peter. 1998. *Infelicities. Representations of the exotic*. Baltimore and London: The Johns Hopkins University Press.

Pomian, Krzysztof. 1987. *Collectionneurs, amateurs et curieux. Paris-Venise: XVIe–XVIIIe siècle*. Paris: Gallimard.

Russo, Alessandra. 2008. "Image-plume, temps reliquaire? Tangibilité d'une histoire esthétique Nouvelle-Espagne, XVIe–XVIIe siecles)". *Images Re-vues* Hors-série 1, 2008, made available online 22 April 2011. http://imagesrevues.revues.org/988 (accessed 30 September 2016).

Schmidt, Benjamin. 2015. *Inventing exoticism. Geography, Globalism, and Europe's Early Modern World*. Philadelphia: University of Pennsylvania Press.

Schnapper, Antoine. 1988. *Le géant, la licorne, la tulipe. Les cabinets de curiosités en France au XVIIe siècle*. Paris: Flammarion.

Swadling, Pamela. 1997. *Plumes from Paradise: trade cycles in Outer Southeast Asia and their impact on New Guinea and nearby islands until 1920*. Boroko: Papua New Guinea National Museum and Robert Browne.

Swan, Claudia. "Exotica on the move. Birds of Paradise in Early Modern Holland." *Art History* 18, vol. 34 (2015). 620–635.

Trichet, Pierre. 1631–1635. *Synopsis rerum variarum, tam naturalium quam artificialium, quae in musaeo Petri Tricheti Burdigalae reperiuntur.—Dénombrement de diverses et curieuses choses du cabinet de Pierre Trichet, bourdelois*. s. l. Paris: Bibliothèque nationale de France.

Index

A

Administration/administrative, 42, 48–62, 99
Alain de Lille, 1
Albertus Magnus, 130, 134–136, 142, 160
Aldrovandi, Ulisse, 184
Allegory/allegoric, 144
Amphibian, 3, 106
Amulet, 105
Ancestor, 113
Animal turn, 104
Antiquity/Antique, 4, 23, 103, 104, 111, 112, 130, 131, 155, 184, 185
Archaeology, vi, 17–19, 96, 100, 102, 104, 113, 123, 163
Aristotle, 74, 84, 121–123, 127, 129, 130, 132, 134–136, 139, 142, 145
Artificialia, 182, 184
Ass, 70, 72, 73, 132
Augustinus Hipponensis, St. Augustine of Hippo, 88
Aurochs, 122, 129, 138
Avar Khaganate, 107, 108, 110, 113, 114
Avicenna, Ibn Sīnā, 90, 91, 134

B

Balkans, 104, 105, 109, 112
Bartholomaeus Anglicus, Bartholomew the Englishman, 82, 83, 89, 91, 92, 128, 130–134, 139, 142–145
Bear, 78, 84, 89–92, 131, 132, 137, 168, 171
Beaver, 84, 88, 89, 172
Bestiary, 3, 69, 71, 122, 123, 126–132, 134–136, 139, 140, 142, 144, 145, 152, 153, 161
Bible/Biblical, 3, 74, 83, 89, 144, 145
Bird of Paradise/*Manucodiata*, 181, 183
Bohemia, 69, 130–134, 137, 139
Bonasus/bonasos, 121–123, 126–128, 130–132, 138, 139, 144, 146

L. Bartosiewicz, A. M. Choyke (eds.), *Medieval Animals on the Move*, https://doi.org/10.1007/978-3-030-63888-7

Bonnacon/bonacon, 124–129, 131, 132, 137, 141, 142, 144, 145
Bovine, 122, 138
Breeding, husbandry, 22, 62
Buffalo, 122, 154
Buffon, Georges-Louis Leclerc, 122, 123
Bull, 4, 122, 127, 135, 136, 142, 146, 173
Burial practice, burial rite, 97, 99, 100, 114
Byzantine/Byzantium, 96, 99, 100, 168, 172, 174

C
Cabinet of curiosities, 7, 179–186
Camel, 44, 173
Caprine, 17, 26, 32, 33
Carnivore, 172
Caspian whipsnake, *Dolichophis caspius*, 105
Cat, 13, 160–162, 168
Caterpillar, 70
Cattle, 8, 13, 22, 29, 33, 45, 138, 168
Centaur, 151
Christianity/Christian, 3, 69, 83, 96, 99–101, 112, 129, 182
Chronicle, 46, 49, 51, 69, 137, 145
Cimex, 158
Cistercian, 74–77, 159
Communication, 42, 51, 52, 105
Composite/hybrid, 70, 122, 123, 182
Cow, 4, 19–21, 23, 135
Crocodile, 153, 154
Cultural deposit, 95, 97, 103, 105, 113
Cultural memory, 104, 113, 123

D
Dante Alighieri, 81, 162, 163
Defense, 136, 142
Devil, 74, 89, 152
Didactic text, 152
Dionysian cult, 105
Disease, 44–47, 54, 59–61, 109
Dog, 3, 13, 69, 84–88, 131, 132, 135–137, 139, 141–143, 145, 152, 168
Domestic, 2, 8, 14, 21, 31–33, 41, 85, 137, 155, 168
Dominican, 81–83, 85, 86, 130, 143
Dragon, 151–154
Dung/excrement, 4, 122, 127, 135, 136, 139–142

E
Eagle, 155, 156, 169
Eckhart, Meister, 78
Eel, 77
Elephant, 3, 154, 171, 174
Elk, 129
Encyclopedia, 42, 44, 81–84, 91, 92, 128, 130, 134
Ermine, 172
Estienne, Charles, 160, 161
Exemplum/exemplar, 3, 87, 88, 92, 143, 144
Exotic, 2, 5, 7, 83, 104, 114, 153–155, 159, 167–175, 183, 184

F
Falcon, 71, 169
Faroe Islands, 32–35
Fish, 4, 14, 19, 24, 30, 32, 70, 74, 83, 157, 159
Flight, 68, 71, 136, 145, 173, 181
Fodder, 30, 33, 35
Four-lined snake, *Elaphe quatuorlineata*, 105
Fox, 68, 109, 110, 169, 171–173
Franciscan, 83
Fume, 127, 140

G
Gaston Phébus, 162
Gervase of Tilbury, 130, 131, 139, 143, 145
Goat, 8, 13–35, 74, 132, 137, 168
Grave, 95–114
Greenland, 4, 5, 7, 32, 34, 35, 169–171
Gregorius Magnus, Saint Gregory the Great, 86, 87
Grub, maggot, 158
Gyrfalcon, 71

H
Hare, 68, 172
Hawk, 71
Horn, 4, 6–9, 26, 122, 127, 131, 132, 134–137, 141, 143, 145, 173, 180
Horse/equine, 2, 13, 41–62, 69, 70, 122, 127, 135, 152, 154, 168, 173
Household, 15, 49, 53, 54, 57–60, 62
Hunting, 3, 104, 129, 145, 160, 170
Hybrid, 123, 182

I
Ibn Fadlan, 173, 174
Iceland, 4, 5, 9, 13–35, 168–171, 174
Imaginary, 3, 4, 151, 153–155
Imperato, Ferrante, 179
Imperial, 41–62
Intertextual variation, 142
Isidorus Hispalensis, Isidore of Seville, Isidore of Sevilla, 68, 70, 72, 88, 158, 161
Ivory/tusk, 7, 32, 35, 170, 171, 186

J
Jordan Catala, 155

K
Knowledge transfer, 45, 48, 62, 108, 138

L
Lamb, 21, 27
Leopard, *Panthera pardus,* pard, 152, 153, 155, 158–163
Leucrota, 70
Lion, 70, 89, 152, 154, 159, 160, 162, 174
Livestock, 2, 13–17, 19, 20, 22, 23, 27, 29–35, 104
Locust, 70
Lynx, 159–163

M
Macaw parrot, 181
Marco Polo, 154, 155
Marten, 172
Meat, 22, 26, 29, 30, 32, 41, 101, 122, 156, 168, 170
Medicinal, 105, 132
Metaphor/metaphoric, 3
Mexico, 182
Military, 41, 42, 48, 49, 51, 53, 62, 99
Milk, 19, 20, 25, 29, 30, 32
Ming Dynasty, 2, 41–44, 48–62
Mink, 172
Mobility, 3, 41, 48, 145
Monapos, 121, 134
Monkey, 174
Moralizing comparison, 71, 72
Mythology/mythological, 168

N
Nanzhili region, 53–61
Narwhal, *Monodon monoceros,* narval, 4–7

Naturalia, 182
New World, 180, 181, 183, 184
Nomad, 42

O
Olaus Magnus, 4
Old Testament, 74
Ontological status, 123
Ostrich, 71
Otto von Guericke, 4
Ounce, 159, 160
Ox, 129, 132, 134–137
Oyster, 156, 157

P
Paeonian, 122, 145
Panther, 152, 153, 155, 159–163
Pastoral, 13, 16
Pasture, 2, 27, 42, 46, 47, 49, 54, 57–59, 74
Peacock, 174
Pelican, 174
Phrygia/Phrygian, 123, 127, 130, 145
Physiologus, 123, 126, 127, 130, 144
Pig, 13–35, 168
Plinius Maior, Pliny the Elder, 87, 90
Poland, 82, 130, 131, 137, 139
Polyceraty, 8
Poultry, 13
Prey, 121–146, 155
Projectile, 121, 131, 134, 136, 139, 141, 142, 146
Ptarmigan, 169, 170
Publishing, 44, 45

R
Ram, 8, 9, 21
Raven, 169
Reindeer, 168
Renaissance, 7, 9, 161
Renart, 162
Reptile, 3
Reykjavík, 16
Rhinoceros, 154, 155, 157, 159
Rodent, 3, 172
Roman, 95–97, 100, 103, 105, 108–113, 184
Rural, 2, 16, 22, 42, 53, 57

S
Sacrificial, 107
Semantic, 153
Semiophore, 182
Sermon, 69, 74, 81–83, 133, 134, 143
Sheep, 8, 9, 13, 17, 20–22, 25–27, 29–34, 168, 173
Similitudo, 88, 92
Siren, 70, 151
Sloth, 72, 73
Snake, serpent, 154
Solinus, 123, 127, 135–137, 139–143, 161
Speed, velocity, 67–71, 74, 78
Squirrel, 76, 168, 172
Supernatural, 23, 27, 144
Swan, 169, 181
Symbol/symbolic, 3, 9, 31, 73, 108, 111, 168
Symon Simeonis, 153, 154

T
Taphonomy/taphonomic, 103, 112
Taxonomy/taxonomic, 4, 151
Tax/taxation, 17, 52–53, 55
Terminology/terminological, 153, 155, 160
Testudo, 156, 157

Thomas of Cantimpré, 68, 70, 74, 130, 134–136, 138, 140, 142, 143, 145
Tiger, 68, 69, 154
Tomaso dei Cavalieri, 184
Totem, 1, 104
Toucan, 181
Trade, 2, 6, 7, 31, 32, 34, 52, 62, 163, 164
Turtle, 155, 156

U
Unicorn, 4, 6, 7, 151, 154, 155, 157, 159, 161, 180

V
Veterinary, 2, 41–62
Viking, 6, 9, 167–175
Vincent de Beauvais, 83
Vulture, 68, 72

W
Walrus, *Odobenus rosmarus*, 5, 32, 35, 170, 171
Welk, 157, 160
Wether, 21
Whale/whaling, 4, 7, 121, 168, 170
Wild boar, 154
Wolf, 3, 161, 168
Wool, 21, 27, 30–33, 35

Z
Zooarchaeology/archaeozoology, 17, 102

The manufacturer's authorised representative in the EU is Springer Nature Customer Service Centre GmbH, Europaplatz 3, 69115 Heidelberg, Germany. If you have any concerns regarding our products, please contact ProductSafety@springernature.com

Printed and bound by CPI Group (UK) Ltd, Croydon, CR0 4YY
30/09/2026
02225458-0001